Flinn Scientific
ChemTopic™ Labs

Biochemistry–The Molecules of Life

Senior Editor

Irene Cesa
Flinn Scientific, Inc.
Batavia, IL

Curriculum Advisory Board

Bob Becker
Kirkwood High School
Kirkwood, MO

Kathleen J. Dombrink
McCluer North High School
Florissant, MO

Robert Lewis
Downers Grove North High School
Downers Grove, IL

John G. Little
St. Mary's High School
Stockton, CA

Lee Marek
Naperville North High School
Naperville, IL

John Mauch
Braintree High School
Braintree, MA

Dave Tanis
Grand Valley State University
Allendale, MI

FLINN SCIENTIFIC INC.
"Your Safer Source for Science Supplies"
P.O. Box 219 • Batavia, IL 60510
1-800-452-1261 • www.flinnsci.com

ISBN 1-877991-88-0

Copyright © 2002 Flinn Scientific, Inc.

Printed in the United States of America.

Table of Contents

Flinn ChemTopic™ Labs Series Preface

Lab Manuals Organized Around Key Content Areas in Chemistry

In conversations with chemistry teachers across the country we have heard a common concern. Teachers are frustrated with their current lab manuals, with experiments that are poorly designed and don't teach core concepts, with procedures that are rigid and inflexible and don't work. Teachers want greater flexibility in their choice of lab activities. As we further listened to experienced master teachers who regularly lead workshops and training seminars, another theme emerged. Master teachers mostly rely on collections of experiments and demonstrations they have put together themselves over the years. Some activities have been passed on like cherished family recipe cards from one teacher to another. Others have been adapted from one format to another to take advantage of new trends in microscale equipment and procedures, technology innovations, and discovery-based learning theory. In all cases the experiments and demonstrations have been fine-tuned based on real classroom experience.

Flinn Scientific has developed a series of lab manuals based on these "cherished recipe cards" of master teachers with proven excellence in both teaching students and training teachers. Created under the direction of an Advisory Board of award-winning chemistry teachers, each lab manual in the Flinn ChemTopic™ Labs series contains 4–6 student-tested experiments that focus on essential concepts and applications in a single content area. Each lab manual also contains 3–5 demonstrations that can be used to illustrate a chemical property, reaction, or relationship and will capture your students' attention. The experiments and demonstrations in the Flinn ChemTopic™ Labs series are enjoyable, highly focused, and will give students a real sense of accomplishment.

Laboratory experiments allow students to experience chemistry by doing chemistry. Experiments have been selected to provide students with a crystal-clear understanding of chemistry concepts and encourage students to think about these concepts critically and analytically. Well-written procedures are guaranteed to work. Reproducible data tables teach students how to organize their data so it is easily analyzed. Comprehensive teacher notes include a master materials list, solution preparation guide, complete sample data, and answers to all questions. Detailed lab hints and teaching tips show you how to conduct the experiment in your lab setting and how to identify student errors and misconceptions before students are led astray.

Chemical demonstrations provide another teaching tool for seeing chemistry in action. Because they are both visual and interactive, demonstrations allow teachers to take students on a journey of observation and understanding. Demonstrations provide additional resources to develop central themes and to magnify the power of observation in the classroom. Demonstrations using discrepant events challenge student misconceptions that must be broken down before new concepts can be learned. Use demonstrations to introduce new ideas, illustrate abstract concepts that cannot be covered in lab experiments, and provide a spark of excitement that will capture student interest and attention.

Safety, flexibility, and choice

Safety always comes first. Depend on Flinn Scientific to give you upfront advice and guidance on all safety and disposal issues. Each activity begins with a description of the hazards involved and the necessary safety precautions to avoid exposure to these hazards. Additional safety, handling, and disposal information is also contained in the teacher notes.

The selection of experiments and demonstrations in each Flinn ChemTopic™ Labs manual gives you the flexibility to choose activities that match the concepts your students need to learn. No single teacher will do all of the experiments and demonstrations with a single class. Some experiments and demonstrations may be more helpful with a beginning-level class, while others may be more suitable with an honors class. All of the experiments and demonstrations have been keyed to national content standards in science education.

Chemistry is an experimental science!

Whether they are practicing key measurement skills or searching for trends in the chemical properties of substances, all students will benefit from the opportunity to discover chemistry by doing chemistry. No matter what chemistry textbook you use in the classroom, Flinn ChemTopic™ Labs will help you give your students the necessary knowledge, skills, attitudes, and values to be successful in chemistry.

About the Curriculum Advisory Board

Flinn Scientific is honored to work with an outstanding group of dedicated chemistry teachers. The members of the Flinn ChemTopic Labs Advisory Board have generously contributed their proven experiments, demonstrations, and teaching tips to create these topic lab manuals. The wisdom, experience, creativity, and insight reflected in their lab activities guarantee that students who perform them will be more successful in learning chemistry. On behalf of all chemistry teachers, we thank the Advisory Board members for their service to the teaching profession and their dedication to the field of chemistry education.

Bob Becker teaches chemistry and AP chemistry at Kirkwood High School in Kirkwood, MO. Bob received his B.A. from Yale University and M.Ed. from Washington University and has 15 years of teaching experience. A well-known demonstrator, Bob has conducted more than 100 demonstration workshops across the U.S. and Canada and is currently a Team Leader for the Flinn Foundation Summer Workshop Program. His creative and unusual demonstrations have been published in the *Journal of Chemical Education,* the *Science Teacher,* and *Chem13 News*. Bob is the author of two books of chemical demonstrations, *Twenty Demonstrations Guaranteed to Knock Your Socks Off, Volumes I and II,* published by Flinn Scientific. Bob has been awarded the James Bryant Conant Award in High School Teaching from the American Chemical Society, the Regional Catalyst Award from the Chemical Manufacturers Association, and the Tandy Technology Scholar Award.

Kathleen J. Dombrink teaches chemistry and advanced-credit college chemistry at McCluer North High School in Florissant, MO. Kathleen received her B.A. in Chemistry from Holy Names College and M.S. in Chemistry from St. Louis University and has more than 30 years of teaching experience. Recognized for her strong support of professional development, Kathleen has been selected to participate in the Fulbright Memorial Fund Teacher Program in Japan and NEWMAST and Dow/NSTA Workshops. She served as co-editor of the inaugural issues of *Chem Matters* and was a Woodrow Wilson National Fellowship Foundation Chemistry Team Member for more than 10 years. Kathleen is currently a Team Leader for the Flinn Foundation Summer Workshop Program. Kathleen has received the Presidential Award, the Midwest Regional Teaching Award from the American Chemical Society, the Tandy Technology Scholar Award, and a Regional Catalyst Award from the Chemical Manufacturers Association.

Robert Lewis teaches chemistry and AP chemistry at Downers Grove North High School in Downers Grove, IL. Robert received his B.A. from North Central College and M.A. from University of the South and has more than 25 years of teaching experience. He was a founding member of Weird Science, a group of chemistry teachers that has traveled throughout the country to stimulate teacher interest and enthusiasm for using demonstrations to teach science. Robert was a Chemistry Team Leader for the Woodrow Wilson National Fellowship Foundation and is currently a Team Leader for the Flinn Foundation Summer Workshop Program. Robert has received the Presidential Award, the James Bryant Conant Award in High School Teaching from the American Chemical Society, the Tandy Technology Scholar Award, a Regional Catalyst Award from the Chemical Manufacturers Association, and a Golden Apple Award from the State of Illinois.

John G. Little teaches chemistry and AP chemistry at St. Mary's High School in Stockton, CA. John received his B.S. and M.S. in Chemistry from University of the Pacific and has more than 35 years of teaching experience. Highly respected for his well-designed labs, John is the author of two lab manuals, *Chemistry Microscale Laboratory Manual* (DC Heath), and *Microscale Experiments for General Chemistry* (with Kenneth Williamson, Houghton Mifflin). He is also a contributing author to *Science Explorer* (Prentice Hall) and *World of Chemistry* (McDougal Littell). John served as a Chemistry Team Leader for the Woodrow Wilson National Fellowship Foundation from 1988 to 1997 and is currently a Team Leader for the Flinn Foundation Summer Workshop Program. He has been recognized for his dedicated teaching with the Tandy Technology Scholar Award and the Regional Catalyst Award from the Chemical Manufacturers Association.

Lee Marek teaches chemistry and AP chemistry at Naperville North High School in Naperville, IL. Lee received his B.S. in Chemical Engineering from the University of Illinois and M.S. degrees in both Physics and Chemistry from Roosevelt University. He has more than 30 years of teaching experience and is currently a Team Leader for the Flinn Foundation Summer Workshop Program. His students have won national recognition in the International Chemistry Olympiad, the Westinghouse Science Talent Search, and the Internet Science and Technology Fair. Lee was a founding member of ChemWest, a regional chemistry teachers alliance, and led this group for 14 years. Together with two other ChemWest members, Lee also founded Weird Science and has presented 500 demonstration and teaching workshops for more than 300,000 students and teachers across the country. Lee has performed science demonstrations on the *David Letterman Show* 20 times. Lee has received the Presidential Award, the James Bryant Conant Award in High School Teaching from the American Chemical Society, the National Catalyst Award from the Chemical Manufacturers Association, and the Tandy Technology Scholar Award.

John Mauch teaches chemistry and AP chemistry at Braintree High School in Braintree, MA. John received his B.A. in Chemistry from Whitworth College and M.A. in Curriculum and Education from Washington State University and has 25 years of teaching experience. John is an expert in "writing to learn" in the chemistry curriculum and in microscale chemistry. He is the author of two lab manuals, *Chemistry in Microscale, Volumes I and II* (Kendall/Hunt). He is also a dynamic and prolific demonstrator and workshop leader. John has presented the Flinn Scientific Chem Demo Extravaganza show at NSTA conventions for seven years and has conducted more than 100 workshops across the country. John was a Chemistry Team Member for the Woodrow Wilson National Fellowship Foundation program for four years and is currently a Team Leader for the Flinn Foundation Summer Workshop Program.

Dave Tanis is Associate Professor of Chemistry at Grand Valley State University in Allendale, MI. Dave received his B.S. in Physics and Mathematics from Calvin College and M.S. in Chemistry from Case Western Reserve University. He taught high school chemistry for 25 years before joining the staff at Grand Valley State University to direct a coalition for improving pre-college math and science education. Dave later joined the faculty at Grand Valley State University and currently teaches courses for pre-service teachers. The author of two laboratory manuals, Dave acknowledges the influence of early encounters with Hubert Alyea, Marge Gardner, Henry Heikkinen, and Bassam Shakhashiri in stimulating his long-standing interest in chemical demonstrations and experiments. Continuing this tradition of mentorship, Dave has led more than 40 one-week institutes for chemistry teachers and served as a Team Member for the Woodrow Wilson National Fellowship Foundation for 13 years. He is currently a Board Member for the Flinn Foundation Summer Workshop Program. Dave received the College Science Teacher of the Year Award from the Michigan Science Teachers Association.

Preface
Biochemistry–The Molecules of Life

Carbohydrates, proteins, lipids, and nucleic acids—these are the molecules of life. From the smallest microbe to the largest mammal, all life depends on the properties and reactions of these four classes of organic compounds. Carbohydrates provide energy for metabolism. Proteins catalyze the reactions that allow an organism to live and grow. Lipids build cell membranes that regulate the internal structure and function of cells. Nucleic acids supply the blueprint for the synthesis of proteins and the function of life itself. Biochemistry is the study of the structure, properties, and reactions of these compounds. The purpose of *Biochemistry,* Volume 20 in the Flinn ChemTopic™ Labs series, is to bring together a selection of biochemistry lab activities for the high school science classroom. Four experiments and five demonstrations allow students to explore the physical and chemical properties of biological compounds and build vital connections between biology and chemistry.

Carbohydrates

Carbohydrates are the most abundant biological compounds. Although all carbohydrates are related structurally, they also have essential differences that give them characteristic chemical properties. In "Introduction to Carbohydrates," students perform a sequence of classification tests to unscramble the identities of five unknown carbohydrates. How are monosaccharides related to di- and polysaccharides? What is the difference between a reducing and nonreducing sugar? Each classification test addresses a key question concerning the structure and properties of carbohydrates and provides a single clue that will reveal the identity of an unknown. Three demonstrations illustrate the connection between the physical and chemical properties of carbohydrates and their biological roles. "Membrane Diffusion" compares the diffusion of small and large molecules across a semipermeable membrane to mimic the process of diffusion in cells. "Glucose Fermentation" uses an acid–base indicator and a redox indicator to demonstrate the chemistry involved in fermentation. Finally, the role of enzymes in carbohydrate digestion is explored in "Lactose Intolerance."

Amino acids and proteins

Starting with only 20 different amino acids, a single cell may synthesize more than 3,000 different proteins. The question naturally arises—what are the similarities and differences among all these different proteins? In "Identifying Proteins and Amino Acids," students perform a series of chemical tests to investigate the structure and composition of proteins and amino acids. The similarities among different proteins—albumin, casein, and gelatin—are first revealed using the biuret test. Their differences are then explored using chemical tests for three specific amino acid residues. The identification of amino acids and proteins is also the basis of an applications-oriented demonstration in "Amino Acid Fingerprints," which looks at the forensic use of the ninhydrin reaction to detect latent fingerprints. In a second experiment, "Physical Properties of Proteins," students identify possible physiological and environmental factors that lead to protein denaturation and loss of protein activity. The effect of pH on the structure and properties of proteins is further illustrated in "pH and Protein Solubility—A Reversible Demonstration."

Lipids

The chemistry of lipids is more varied than that of either carbohydrates or proteins. Lipids are related to each other not by a common structure but by a common physical property—they are insoluble in water. In "Properties of Lipids," students study the real-life processes that are used to obtain and characterize lipids. Students also perform a test for unsaturation on a variety of seed oils to learn more about the vital role of saturated versus unsaturated fats in nutrition and health. By comparing their results with the information provided on the nutritional labels of different foods, students gain personal insight into the connections between biology, chemistry, and nutrition.

Building connections

The selection of experiments and demonstrations in *Biochemistry— The Molecules of Life* provides an excellent introduction to the chemical basis of life. Building connections between the sciences and integrating all aspects of science content are important goals of science education, and indeed these goals are formally embedded in the National Science Education Standards. The experiments in this book allow students to build essential connections linking previous knowledge in biology, chemistry, and nutrition. The demonstrations add further perspective to tie together the biological and chemical properties of organic compounds. Each experiment and demonstration in *Biochemistry—The Molecules of Life* has been thoroughly tested and retested to assure success. Use the experiment summaries and concepts on the following pages to locate the concepts you want to teach and to choose experiments and demonstrations that will help you meet your goals.

Format and Features

Flinn ChemTopic™ Labs

All experiments and demonstrations in Flinn ChemTopic™ Labs are printed in a $10\frac{7}{8}'' \times 11''$ format with a wide 2″ margin on the inside of each page. This reduces the printed area of each page to a standard $8\frac{1}{2}'' \times 11''$ format suitable for copying.

The wide margin assures you the entire printed area can be easily reproduced without hurting the binding. The margin also provides a convenient place for teachers to add their own notes.

Concepts

Use these bulleted lists along with state and local standards, lesson plans, and your textbook to identify activities that will allow you to accomplish specific learning goals and objectives.

Background

A balanced source of information for students to understand why they are doing an experiment, what they are doing, and the types of questions the activity is designed to answer. This section is not meant to be exhaustive or to replace the students' textbook, but rather to identify the core concepts that should be covered before starting the lab.

Experiment Overview

Clearly defines the purpose of each experiment and how students will achieve this goal. Performing an experiment without a purpose is like getting travel directions without knowing your destination. It doesn't work, especially if you run into a roadblock and need to take a detour!

Pre-Lab Questions

Making sure that students are prepared for lab is the single most important element of lab safety. Pre-lab questions introduce new ideas or concepts, review key calculations, and reinforce safety recommendations. The pre-lab questions may be assigned as homework in preparation for lab or they may be used as the basis of a cooperative class activity before lab.

Materials

Lists chemical names, formulas, and amounts for all reagents—along with specific glassware and equipment—needed to perform the experiment as written. The material dispensing area is a main source of student delay, congestion, and accidents. Three dispensing stations per room are optimum for a class of 24 students working in pairs. To safely substitute different items for any of the recommended materials, refer to the *Lab Hints* section in each experiment or demonstration.

Safety Precautions

Instruct and warn students of the hazards associated with the materials or procedure and give specific recommendations and precautions to protect students from these hazards. Please review this section with students before beginning each experiment.

Procedure

This section contains a stepwise, easy-to-follow procedure, where each step generally refers to one action item. Contains reminders about safety and recording data where appropriate. For inquiry-based experiments the procedure may restate the experiment objective and give general guidelines for accomplishing this goal.

Data Tables

Data tables are included for each experiment and are referred to in the procedure. These are provided for convenience and to teach students the importance of keeping their data organized in order to analyze it. To encourage more student involvement, many teachers prefer to have students prepare their own data tables. This is an excellent pre-lab preparation activity—it ensures that students have read the procedure and are prepared for lab.

Post-Lab Questions or Data Analysis

This section takes students step-by-step through what they did, what they observed, and what it means. Meaningful questions encourage analysis and promote critical thinking skills. Where students need to perform calculations or graph data to analyze the results, these steps are also laid out sequentially and in order.

Format and Features

Teacher's Notes

Master Materials List

Lists the chemicals, glassware, and equipment needed to perform the experiment. All amounts have been calculated for a class of 30 students working in pairs. For smaller or larger class sizes or different working group sizes, please adjust the amounts proportionately.

Preparation of Solutions

Calculations and procedures are given for preparing all solutions, based on a class size of 30 students working in pairs. With the exception of particularly hazardous materials, the solution amounts generally include 10% extra to account for spillage and waste. Solution volumes may be rounded to convenient glassware sizes (100 mL, 250 mL, 500 mL, etc.).

Safety Precautions

Repeats the safety precautions given to the students and includes more detailed information relating to safety and handling of chemicals and glassware. Refers to Material Safety Data Sheets that should be available for all chemicals used in the laboratory.

Disposal

Refers to the current *Flinn Scientific Catalog/Reference Manual* for general guidelines and specific procedures governing the disposal of laboratory waste. Because we recommend that teachers review local regulations before beginning any disposal procedure, the information given in this section is for general reference purposes only. However, if a disposal step is included as part of the experimental procedure itself, then the specific solutions needed for disposal are described in this section.

Lab Hints

This section reveals common sources of student errors and misconceptions and where students are likely to need help. Identifies the recommended length of time needed to perform each experiment, suggests alternative chemicals and equipment that may be used, and reminds teachers about new techniques (filtration, pipeting, etc.) that should be reviewed prior to lab.

Teaching Tips

This section puts the experiment in perspective so that teachers can judge in more detail how and where a particular experiment will fit into their curriculum. Identifies the working assumptions about what students need to know in order to perform the experiment and answer the questions. Highlights historical background and applications-oriented information that may be of interest to students.

Sample Data

Complete, actual sample data obtained by performing the experiment exactly as written is included for each experiment. Student data will vary.

Answers to All Questions

Representative or typical answers to all questions. Includes sample calculations and graphs for all data analysis questions. Information of special interest to teachers only in this section is identified by the heading "Note to the teacher." Student answers will vary.

Look for these icons in the *Experiment Summaries and Concepts* section and in the *Teacher's Notes* of individual experiments to identify inquiry-, microscale-, and technology-based experiments, respectively.

Experiment Summaries and Concepts

Experiment

Introduction to Carbohydrates—Structure and Properties

What is a carbohydrate? How are monosaccharides related to di- and polysaccharides? What is the difference between a reducing and nonreducing sugar? The purpose of this experiment is to explore the structures and properties of different types of carbohydrates and learn how they can be identified. The identities of five carbohydrates—starch, glucose, fructose, lactose, and sucrose—have been scrambled. Students perform a set of classification tests in sequence to unscramble the carbohydrate code and reveal the identities of the unknowns.

Identifying Proteins and Amino Acids

Starting with only about 20 different amino acids, a single cell may synthesize more than 3,000 different types of proteins. How do amino acids link together to build a protein? The purpose of this experiment is to identify proteins and amino acids using a series of classification tests. Students study the behavior of albumin, casein, and gelatin using the biuret test to identify the general nature of the peptide linkage and its central role in protein structure. A series of chemical tests are then performed to investigate the composition of these proteins with respect to three specific amino acids.

Physical Properties of Proteins

The relationship between structure and function is a universal theme in biochemistry. In this experiment, students examine the effects of acids and bases, inorganic salts, organic solvents, and heat on the physical properties of proteins. The results demonstrate how sensitive proteins are to physiological conditions and to the addition of external agents. Many factors are shown to affect the physical properties of proteins that relate, in turn, to their structure and function. Students learn about the interactions among amino acid side chains that allow proteins to fulfill their vital biological functions.

Properties of Lipids

Fats and oils, waxes and cholesterol, steroid hormones and Vitamin A—all of these natural products belong to the diverse class of biological compounds called lipids. The purpose of this experiment is to identify and classify lipids and examine their properties. Students study the solubility of lipids, stain them using a special "fat stain," and perform a chemical test to classify oils as saturated versus unsaturated. Finally, students extract peanut oil from peanuts to determine the amount of fat in peanuts. This activity mimics the real-life processes used to characterize lipids and allows students to compare their results with information provided on the nutritional labels of a variety of food items.

Concepts

- Carbohydrates
- Monosaccharide
- Disaccharide
- Polysaccharide
- Reducing sugars

- Proteins
- Amino acids
- Peptide linkage
- Biuret test
- Xanthoproteic test

- Protein folding
- Native structure
- Denaturation
- Salting-out

- Lipids
- Polar vs. nonpolar
- Triglycerides
- Fats and oils
- Saturated vs. unsaturated

Experiment Summaries and Concepts

Demonstration

Concepts

Membrane Diffusion—Dialysis Demonstration

How does the membrane around a cell help to regulate the internal makeup of the cell? This demonstration compares the diffusion of small and large molecules across a semipermeable membrane to illustrate the process of diffusion in cells. Dialysis tubing is used to model the behavior of a cell membranes with respect to the sizes of the molecules that will or will not diffuse through them. Chemical tests reveal that sodium chloride is small enough to pass through the membrane, but that starch is too large.

- Semipermeable membrane
- Diffusion
- Dialysis

Glucose Fermentation—Metabolism Demonstration

The overall products of the fermentation of glucose, the main carbohydrate in fruits and grains, are ethyl alcohol and carbon dioxide. Many different intermediate products may also be formed depending on reaction conditions. In this demonstration, an acid–base indicator is used to detect the production of carbon dioxide, and a redox indicator is used to illustrate the changing reaction conditions during the fermentation process. An optional distillation procedure is also provided to identify ethyl alcohol in the product mixture.

- Carbohydrates
- Glucose
- Fermentation

Lactose Intolerance—Enzyme Digestion Demonstration

Many people are said to be "lactose intolerant." These people are unable to digest milk and dairy products because they lack the enzyme required to break the linkage joining the two monosaccharide units in lactose (milk sugar). In this demonstration, yeast is used as a model lactose-intolerant organism to illustrate the use of a commercial enzyme product called Lactaid™ in milk digestion.

- Lactaid™
- Lactose
- Disaccharide
- Monosaccharide

Amino Acid Fingerprints—Ninhydrin Demonstration

Latent fingerprints are composed of several chemicals that are naturally present in skin oils or released through the pores of the skin via perspiration. Both amino acids and peptides are normally found in the natural oils on skin. These compounds react with a special reagent, called ninhydrin, to give a characteristic purple product. Detectives use ninhydrin to reveal fingerprints left behind at crime scenes. This demonstration simulates the use of ninhydrin in forensic chemistry to detect latent fingerprints on porous surfaces, such as paper and cloth.

- Amino acids
- Ninhydrin
- Forensic chemistry

pH and Protein Solubility—A Reversible Demonstration

Any changes in the pH of a protein's environment will cause observable changes in the solubility of the protein. These changes occur because proteins contain many acidic and basic groups that have different structures at different pH values. This demonstration looks at the reversible solubility behavior of casein, the principal protein in milk, as hydrochloric acid or sodium hydroxide is added to the protein. The structure of the protein at different pH values is discussed.

- Proteins
- Isoelectric point
- Solubility
- pH

Teacher Notes

Introduction to Carbohydrates
Structure and Properties

Introduction

What is a carbohydrate? What are the roles of carbohydrates in energy, metabolism, and cell structure? Let's explore the structure and properties of different types of carbohydrates and learn how they can be identified.

Concepts

- Carbohydrates
- Monosaccharides
- Disaccharides
- Polysaccharides
- Classification tests
- Reducing vs. nonreducing sugars

Background

Carbohydrates are the most abundant biological compounds. It is estimated that more than 50% of the total carbon content on the Earth is present in the form of carbohydrate compounds. The term carbohydrate dates back to the 1800s, when it was found that the formulas of the simplest carbohydrates could be expressed in the form $(CH_2O)_n$. Both glucose and fructose, for example, have the formula $C_6H_{12}O_6$ and can be written as $(CH_2O)_6$. Carbohydrates appeared to consist of "hydrates" of carbon $(C \cdot H_2O)_n$. Carbohydrates are also referred to as sugars, a common name that reflects the fact that many carbohydrates are naturally occurring sweeteners.

The biological properties of carbohydrates are usually divided into two categories. Many carbohydrates play a primary role in meeting the energy requirements of cells, either as an immediate source of energy or as a means of storing energy for future use. Other carbohydrates serve a structural role within cells—carbohydrates are the primary components of the cell wall in both plants and bacteria.

The simplest carbohydrates are called *monosaccharides*. The two most common monosaccharides are glucose (also called dextrose) and fructose. Monosaccharides are the fundamental units, or building blocks, that make up all other carbohydrates. When two monosaccharide units are joined together, they form *disaccharides*. Examples of disaccharides include sucrose and lactose. *Polysaccharides* are huge organic molecules—called polymers—composed of hundreds or thousands of monosaccharides joined together. The most familiar polysaccharides are starch and cellulose.

Glucose ("blood sugar") is the most abundant monosaccharide in the human body. It is the chemical "fuel" that is carried in the bloodstream to tissues as an energy source for metabolism. Other carbohydrates that are absorbed by the human body must be converted to glucose prior to metabolism. Fructose is the most abundant carbohydrate in fruits. Honey is a 1:1 mixture of glucose and fructose. Although glucose and fructose share the same molecular formula, their structures are different and they have different properties. Fructose is the sweetest sugar—about 30% sweeter per gram than table sugar—and is widely used as a lower calorie and lower cost sweetener than table sugar. Many soft drinks and fruit drinks are sweetened with "high fructose corn syrup."

See the Supplementary Information *section (page 12) for the structures of key carbohydrates.*

Introduction to Carbohydrates

Sucrose, or table sugar, is a disaccharide composed of a glucose unit and a fructose unit joined together. It is harvested from sugar cane or sugar beets and is also called cane sugar. It is a widely used refined sugar in a typical Western diet. Lactose is a disaccharide composed of glucose and a second monosaccharide called galactose. Lactose ("milk sugar") constitutes about 4–5% of cow's milk and 7–8% of human breast milk. A specific enzyme called lactase is required for the digestion of lactose (to break the linkage joining the glucose and galactose units). People who lack this enzyme are said to be lactose intolerant—they cannot digest milk or milk products.

The polysaccharide starch is composed of thousands of glucose units joined together. Found in storage vacuoles in plants, starch is the principal way in which plants store chemical energy.

Classification Tests

The structural and chemical differences among the carbohydrates are the basis of a series of *classification tests* that have been developed to identify carbohydrates in the lab.

Reaction with iodine is used as a classification test to identify the polysaccharide starch. Starch binds iodine molecules to form dark blue–colored complexes. Other carbohydrates do not react with iodine.

Benedict's test is a classification test that is used to identify *reducing sugars,* which include **all** monosaccharides and most disaccharides (excluding sucrose). In contrast, all polysaccharides are *nonreducing sugars*. Benedict's solution contains copper(II) sulfate dissolved in strong base. Cu^{2+} ions in Benedict's solution are a strong oxidizing agent. The Cu^{2+} ions oxidize carbohydrate molecules, and are themselves reduced in the process to copper(I). This is the origin of the term reducing sugar. A positive Benedict's test result is marked by the disappearance of the blue color due to Cu^{2+} ions and the appearance of a red precipitate consisting of reduced copper(I) oxide (Cu_2O). Not all disaccharides will reduce Benedict's solution. The most common nonreducing disaccharide is sucrose.

Benedict's solution gives positive test results with all reducing sugars. Not all reducing sugars react at the same rate, however, with different oxidizing agents—disaccharides are considerably less reactive than monosaccharides. Reducing sugars can be classified as mono- versus disaccharides by reacting them with Barfoed's solution, a weaker oxidizing agent containing copper(II) acetate in acetic acid. A positive Barfoed's test result is similar to that observed with Benedict's solution. Monosaccharides give positive Barfoed's test results within 2–3 minutes, while disaccharides do not react under the same conditions.

The Seliwanoff test is used to distinguish between different types of monosaccharides, specifically, aldoses and ketoses. Glucose and fructose have the same formula, $C_6H_{12}O_6$. They differ only in the nature of one organic functional group they contain.

Glucose is an aldose—it contains an aldehyde functional group ($-\overset{\overset{\displaystyle O}{\|}}{C}-H$ set of atoms).

Fructose is a ketose—it contains a ketone functional group ($-\overset{\overset{\displaystyle O}{\|}}{C}-$ set of atoms).

Teacher Notes

Ketoses readily lose water upon heating with 3 M hydrochloric acid for 2–3 minutes. The resulting compounds react with another reagent, called resorcinol, in the Seliwanoff test to form red products. Aldoses do not react under the same conditions. A color change from colorless to red in the Seliwanoff test serves as a positive result to identify ketoses, such as fructose.

Experiment Overview

Five carbohydrate samples, labeled A–E, are provided. The identities of these carbohydrate samples, starch, glucose, fructose, lactose, and sucrose, have been scrambled. The carbohydrate "code" can be unscrambled by performing four classification tests in sequence. A "blank" sample—distilled water, which always gives negative test results—will be included in each test in order to tell the difference between a positive and a negative test result. As each classification test is performed in sequence (see the *Pre-Lab Questions*), the identity of one of the unknown samples should become known. This sample is then removed from the number of samples that must be carried over to the next classification test in the sequence. Thus, the first test (iodine), must be carried out on six samples (five carbohydrate unknowns plus the blank), but the final test should only need to be run on three samples (the last two carbohydrate unknowns and a distilled water blank).

Pre-Lab Questions

Complete the following Flow Chart to show how the identities of the unknown carbohydrate samples will be revealed using a sequence of four classification tests.

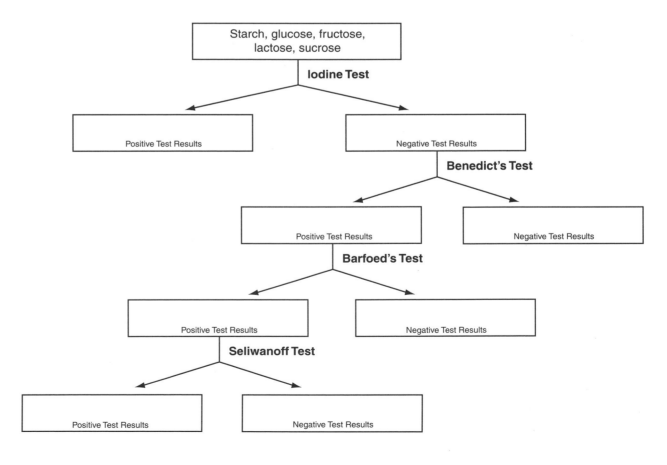

The classification tests must be performed in the sequence shown in the Pre-Lab *section in order to deduce the structures as described in this experiment.*

Materials

Carbohydrate unknowns, A–E

Benedict's solution, 10 mL

Barfoed's solution, 12 mL

Iodine solution, I₂/KI, 2 mL

Seliwanoff reagent, 0.1% resorcinol in 3 M HCl, 6 mL

Water, distilled or deionized

Beaker, 250-mL

Beral-type pipets, 10

Boiling stone

Hot plate

Test tubes, 13×100 mm, 6

Test tube clamp

Test tube rack

Safety Precautions

Iodine solution contains iodine and potassium iodide and is an eye and skin irritant. Benedict's solution contains copper sulfate, sodium citrate, and sodium carbonate; it is moderately toxic by ingestion and a skin and body tissue irritant. Barfoed's solution contains copper acetate and acetic acid; it is moderately toxic by ingestion and a skin and body tissue irritant. The Seliwanoff reagent consists of resorcinol, which is toxic by ingestion, in hydrochloric acid. It is a corrosive liquid. Avoid exposure of all chemicals to eyes and skin. Wear chemical splash goggles, chemical-resistant gloves, and a chemical-resistant apron. Wash hands thoroughly with soap and water before leaving the laboratory.

Procedure

Before beginning Part A, prepare a boiling water bath for use in Parts B–D. Fill a 250-mL beaker half full with water. Add a boiling stone and heat the water to a gentle boil on a hot plate.

Part A. Iodine Test

1. Obtain and label 6 test tubes for the five unknown carbohydrates A–E plus a blank (distilled water).

2. Add 1 mL of each sample to be tested to the correspondingly labeled test tube.

3. Add 2–3 drops of iodine solution to each test tube. Record the color of each solution in the Data Table. Note whether each result is positive or negative.

4. In the Data Table, record the identity of the unknown that has been revealed using this classification test. This unknown can be removed from further testing in Parts B–D.

5. Dispose of the test tube contents in the appropriate iodine waste container, as directed by the instructor. Wash the test tubes and rinse them with distilled water.

Part B. Benedict's Test

6. Label 5 test tubes as needed with the letters of the samples remaining to be tested, including a blank.

7. Place 1 mL of each sample to be tested in the correspondingly labeled test tube.

8. Add 2 mL of Benedict's solution to each test tube and place the test tubes in the boiling water bath.

Remind students of the purpose of the distilled water blank. Water should give negative results (no reaction) with all the tests. Any sample that gives a result different from the blank should be counted as a positive test result.

9. After 2–3 minutes, remove the test tubes from the bath using a test tube clamp. Record the color and appearance of each sample in the Data Table. Note whether each result is positive or negative for the presence of a reducing sugar.

10. In the Data Table, record the identity of the unknown that has been revealed using this classification test. Remember, this unknown can be removed from further testing in Parts C and D.

11. Dispose of the test tube contents down the drain with plenty of water. Wash the test tubes and rinse them with distilled water.

Part C. Barfoed's Test

12. Label 4 test tubes as needed with the letters of the samples remaining to be tested, including a blank.

13. Place 1 mL of each sample to be tested in the correspondingly labeled test tube.

14. Add 3 mL of Barfoed's Reagent to each test tube and place the test tubes in the boiling water bath.

15. After exactly 2 minutes, remove the test tubes from the bath using a test tube clamp. Record the color and appearance of each sample in the Data Table. Note whether each result is positive or negative for the presence of a monosaccharide.

16. In the Data Table, record the identity of the unknown that has been revealed using this classification test. Remember, this unknown can be removed from further testing in Part D.

17. Dispose of the test tube contents down the drain with plenty of water. Wash the test tubes and rinse them with distilled water.

Part D. Seliwanoff Test

18. Label 3 test tubes as needed with the letters of the samples remaining to be tested, including a blank.

19. Place 2 mL of the Seliwanoff reagent in each test tube.

20. Add 2 drops of each sample to be tested in the correspondingly labeled test tube and place the test tubes in the boiling water bath.

21. After 5–6 minutes, remove the test tubes from the bath using a test tube clamp. Record the color of each sample in the Data Table. Note whether each result is positive or negative for the presence of a ketose.

22. In the Data Table, record the identity of the unknown that has been revealed using this classification test.

23. The final unknown can be identified from the list after the first four have been eliminated. Record the identity of the final unknown in the Data Table.

24. Dispose of the test tube contents in the appropriate waste container, as directed by the instructor. Wash the test tubes and rinse them with distilled water.

Follow the recommended time carefully in Part C to avoid a false positive result for lactose.

Name: _____

Class/Lab Period: _____

Introduction to Carbohydrates

Data Table

Sample	Iodine Test		Benedict's Test		Barfoed's Test		Seliwanoff Test		Identity
	Color	+/–	Color	+/–	Color	+/–	Color	+/–	
Blank									Distilled water
A									
B									
C									
D									
E									

Note: Once the identity of an unknown has been revealed, shade the remaining test boxes with hash marks. This will remind you that this unknown has been removed from further testing.

Post-Lab Questions *(Use a separate sheet of paper to answer the following questions.)*

1. Maltose, a product of partial digestion of starches, is a disaccharide composed of two glucose units. It is a reducing sugar. In your mind, take maltose through the sequence of classification tests used in this experiment. Would you be able to distinguish maltose from lactose in an unknown sample?

2. When disaccharides are heated in water in the presence of a strong acid, the linkage joining the two monosaccharide components is "broken" (a reaction called hydrolysis). Using this information, explain why sucrose might give a false positive result with the Seliwanoff reagent.

3. Prior to the advent of more accurate test-strip methods to analyze the amount of glucose in urine (a test for diabetes), the presence of glucose in urine was routinely detected using Clinitest™ tablets. These tablets contain all of the solid reagents needed for Benedict's Test. Do you think this is a good method of testing for glucose?

4. In the Background Section, it was stated that fructose is used as a "lower calorie and lower cost sweetener than table sugar." Explain how and why this statement might be true.

If the tests are performed out of sequence, students may not be able to identify one carbohydrate unambiguously using each test result.

Teacher's Notes
Introduction to Carbohydrates

Master Materials List *(for a class of 30 students working in pairs)*

Carbohydrate Solutions

Fructose, $C_6H_{12}O_6$, 1 g*

Glucose, $C_6H_{12}O_6 \cdot H_2O$, 1 g*

Lactose, $C_{12}H_{22}O_{11} \cdot H_2O$, 1 g*

Starch, 1 g*

Sucrose, $C_{12}H_{22}O_{11}$, 1 g*

Test Solutions

Benedict's solution, 200 mL[†]

Barfoed's solution, 250 mL[†]

Iodine solution, I_2/KI, 50 mL

Seliwanoff reagent, 0.1% resorcinol in 3 M HCl, 125 mL

Equipment

Boiling stones, 10 g

Beakers, 250-mL, 15

Hot plates (or Bunsen burners with ring stand and ring)

Pipets, Beral-type, graduated, 150

Test tube clamps, 15

Test tube racks, 15

Test tubes, borosilicate glass, 13 × 100 mm, 90

Wash bottles, 15

Water, distilled or deionized

*Carbohydrate solutions should be freshly prepared. See the *Preparation of Solutions* section.

[†]We recommend the purchase of ready-made Benedict's (Flinn Catalog No. B0015) and Barfoed's solutions (Flinn Catalog No. B0061).

Preparation of Solutions *(for a class of 30 students working in pairs)*

Carbohydrate Solutions, 1%: Prepare 1% carbohydrate solutions by adding 100 mL of distilled water to 1 g of each carbohydrate. Randomly assign the unknown letter codes A–E to these samples. Place one "unknown" label on each carbohydrate solution. Be sure to record the assignments in your notes before placing the corresponding "unknown" label on the bottle! Prepare these solutions within one week of use.

Iodine/Potassium Iodide Solution: Dissolve 0.75 g of potassium iodide in 15 mL of distilled or deionized water. Add 0.15 g of iodine. Stir to dissolve, then dilute to 50 mL. Store in a dark bottle to prevent exposure to light. Alternatively, purchase Iodine–Potassium Iodide solution (Flinn Catalog No. I0038).

Seliwanoff Reagent: Prepare 125 mL of 3 M hydrochloric acid by diluting 31 mL of concentrated (12 M) HCl to 125 mL. *Note:* Always add acid to water. Allow to cool, then add 0.13 g of resorcinol. Stir to dissolve.

Safety Precautions

Iodine solution contains iodine and potassium iodide and is an eye and skin irritant. Benedict's solution contains copper sulfate, sodium citrate, and sodium carbonate; it is moderately toxic by ingestion and a skin and body tissue irritant. Barfoed's solution contains copper acetate and acetic acid; it is moderately toxic by ingestion and a skin and body tissue irritant. The Seliwanoff reagent consists of resorcinol, which is toxic by ingestion, in hydrochloric acid. It is a corrosive liquid. Avoid exposure of all chemicals to eyes and skin.

Other carbohydrates may also be used as unknowns. Galactose is a monosaccharide that gives identical results to glucose. Maltose is a disaccharide that gives identical results to lactose.

Wear chemical splash goggles, chemical-resistant gloves, and a chemical-resistant apron. Consult current Material Safety Data Sheets for additional safety, handling, and disposal information.

Disposal

Consult your current *Flinn Scientific Catalog/Reference Manual* for both general guidelines and specific methods governing the disposal of laboratory waste. All carbohydrate solutions may be flushed down the drain with excess water according to Flinn Suggested Disposal Method #26b. The waste solutions remaining after each classification test should be collected in separate containers. The waste iodine test solutions can be disposed of according to Flinn Suggested Disposal Method #12a. The waste Benedict's and Barfoed's test solutions may be flushed down the drain with excess water according to Flinn Suggested Disposal Method #26b. Waste Seliwanoff test solutions can be disposed of by neutralization according to Flinn Suggested Disposal Method #24b.

Lab Hints

- The experimental work for this lab can reasonably be completed in one 50-minute lab period.

- The *Pre-Lab* section is included to allow students to determine which unknown carbohydrate is revealed using each of the four classification tests. Students need this information to follow the procedure and to assign the identities of the unknown carbohydrates. The *Pre-Lab Questions* may be assigned as homework in preparation for the lab or may be used as the basis of a cooperative class discussion prior to lab.

- To avoid cross-contamination of samples, consider dispensing beforehand smaller amounts of each solution for each lab bench to use separately. To avoid students sharing their results, provide different unknowns for each group.

- Positive results for Benedict's test range from green to yellow to orange in color, depending on the nature of the carbohydrate, the temperature of the bath, and the reaction time. In each case, however, a reddish precipitate will be evident. Negative test results do not contain a precipitate.

- Both the Barfoed and the Seliwanoff tests require students to time the reactions. The specified times must be carefully followed in order to avoid getting false positive results for disaccharides, in the case of Barfoed's test, and aldoses, in the case of the Seliwanoff test.

- As an extension, consider testing common foods for the presence of specific carbohydrates. The volume of each test solution is sufficient to allow each pair of students to take one food sample through the entire sequence of classification tests, if desired. Samples that might be tested include milk, fruit and vegetable juices (choose light-colored ones, such as apple juice), soda pop (again, colorless is better), and honey or syrup. All of these are more concentrated than the carbohydrate solutions tested in this activity. They should be diluted prior to testing—a good rule of thumb is 1 mL diluted to 50 mL with water. Solid foods—such as crackers, cereals, raw fruits and vegetables—may also be tested by mashing them first in a mortar with pestle and adding water to obtain an "extract."

Teacher Notes

Teaching Tips

- See "Lactose Intolerance" in the *Demonstrations* section of this Flinn ChemTopic™ Labs volume for an interesting demonstration of the metabolism of lactose versus glucose.

- Compare and contrast the structures of starch and cellulose to illustrate how small structural changes can lead to large changes in the biological roles of carbohydrates. Both starch and cellulose consist of thousands of glucose units joined together. The difference between them is the spatial orientation of the linkage joining the monosaccharide units together. This seemingly small structural change makes cellulose totally indigestible to humans! (Termites, on the other hand, thrive by digesting celulose.)

Answers to Pre-Lab Questions *(Student answers will vary.)*

Complete the following Flow Chart to show how the identities of the unknown carbohydrate samples will be revealed using a sequence of four classification tests.

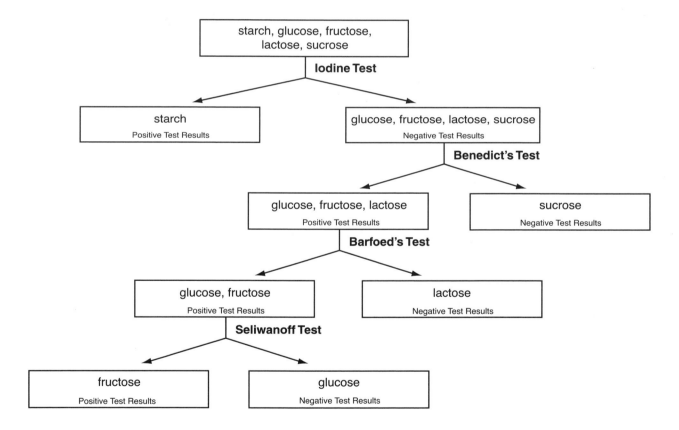

Sample Data

Student data will vary.

Data Table

Sample	Iodine Test		Benedict's Test		Barfoed's Test		Seliwanoff Test		Identity
	Color	+/–	Color	+/–	Color	+/–	Color	+/–	
Blank	yellow-orange	–	blue	–	blue	–	colorless	–	Distilled water
A	yellow-orange	–	red precipitate	+	red precipitate	+	colorless	–	Glucose
B	yellow-orange	–	red precipitate	+	blue	–	///////		Lactose
C	purple	+	///////		///////		///////		Starch
D	yellow-orange	–	blue	–	///////		///////		Sucrose
E	yellow-orange	–	red precipitate	+	red precipitate	+	red	+	Fructose

Note: Carbohydrate identities were assigned as shown in the first and last columns. The samples were removed from further testing once their identities had been revealed by a prior test in the sequence (the test boxes are then shaded with hash marks).

Answers to Post-Lab Questions *(Student answers will vary.)*

1. Maltose, a product of partial digestion of starches, is a disaccharide composed of two glucose units. It is a reducing sugar. In your mind, take maltose through the sequence of classification tests used in this experiment. Would you be able to distinguish maltose from lactose in an unknown sample?

   Maltose would give a negative result in the Iodine test (it is not a starch). As a reducing sugar, it should give a positive result with Benedict's solution. Since it is a disaccharide and not a monosaccharide, however, it gives a negative Barfoed's test. Finally, since maltose is composed only of glucose units joined together, it is an aldose and will give a negative test result with the Seliwanoff reagent. This is the same pattern of results that would be observed with lactose. It is not possible to distinguish maltose and lactose with the classification tests used in this experiment.

2. When disaccharides are heated in water in the presence of a strong acid, the linkage joining the two monosaccharide components is "broken" (a reaction called hydrolysis). Using this information, explain why sucrose might give a false positive result with the Seliwanoff reagent.

   The Seliwanoff reagent contains strong acid (3 M HCl). If a test solution of sucrose were heated with this reagent, the linkage joining the glucose and fructose units would be broken. The fructose would be released into solution and would give a positive test result with the reagent. This is one reason why the directions for running the Seliwanoff test call for heating the solution for a short period of time only (to avoid hydrolysis).

Question #1 suggests an alternative way this experiment could be performed, if desired. Run all the known samples through all four tests, and then compare an unknown against the known samples. Match the pattern of test results to determine the identity of the unknown.

Teacher Notes

3. Prior to the advent of more accurate test-strip methods to analyze the amount of glucose in urine (a test for diabetes), the presence of glucose in urine was routinely detected using Clinitest™ tablets. These tablets contain all of the solid reagents needed for Benedict's test. Do you think this is a good method of testing for glucose?

Benedict's test is not a specific test for glucose. All reducing sugars would give positive test results with Clinitest tablets. Since all monosaccharides and most disaccharides are reducing sugars, the presence of a wide variety of carbohydrates would give false positive readings. Clinitest tablets do provide a valuable general screening tool, however, because the presence of other carbohydrates in urine could be symptoms of other diseases.

4. In the Background Section, it was stated that fructose is used as a "lower calorie and lower cost sweetener than table sugar". Explain how and why this statement might be true.

Since fructose is about 30% sweeter per gram than table sugar, a desired level of sweetness can be achieved using fewer grams of fructose compared to sucrose. Since the calorie content of all sugars is about the same (4 Calories per gram), food sweetened with fructose would supply fewer calories than food sweetened with table sugar. Fructose is a lower cost sweetener than table sugar only if the costs of the two sugars are similar (using fewer grams of fructose costs less than using more grams of table sugar). Of course, all of these observations would no longer be true if more fructose were added to make the foods even sweeter than they would be with added table sugar!

Supplementary Information

Structures of Key Carbohydrates

The structures of glucose, fructose, lactose, sucrose, and starch are shown below. Notice that all the monosaccharide units exist in the form of cyclic, hemiacetal or hemiketal structures. All of the naturally occuring carbohydrates are D-sugars.

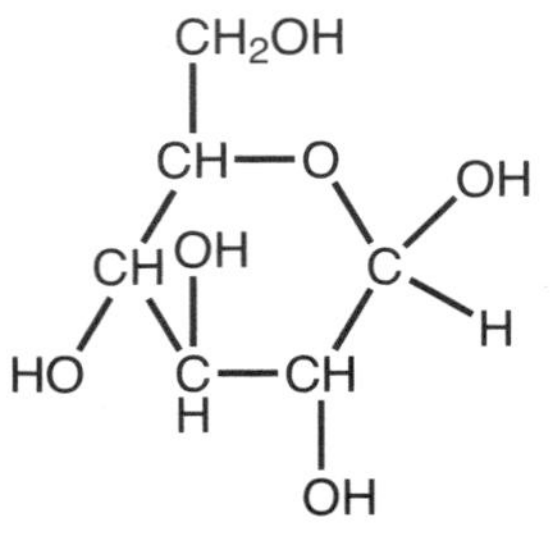

Glucose

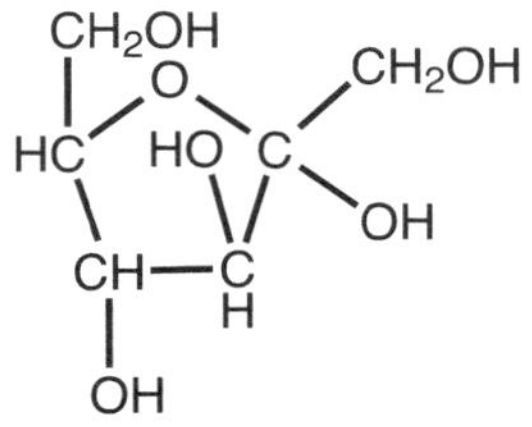

Fructose

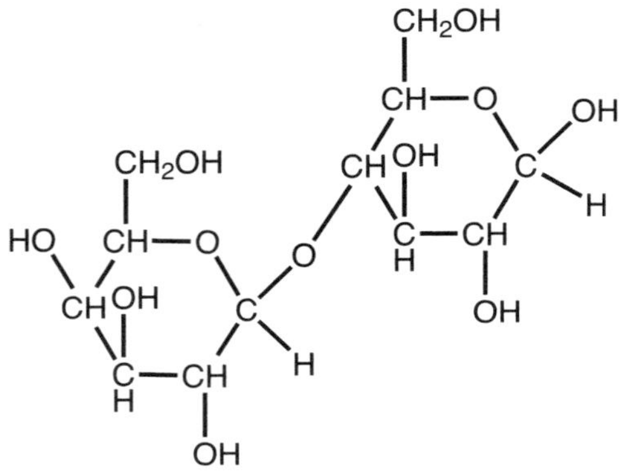

Lactose

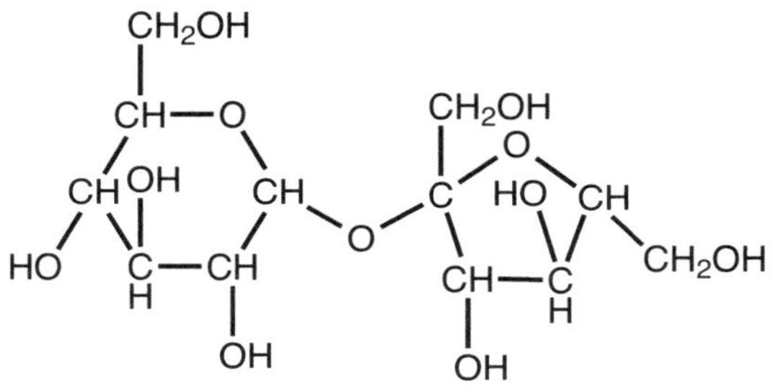

Sucrose

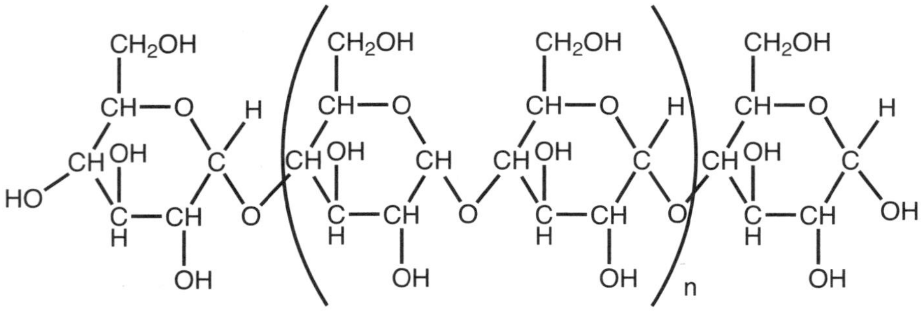

Starch

Identifying Proteins and Amino Acids

Introduction

What are the roles of amino acids in the structure and properties of proteins? Let's investigate the properties of proteins and amino acids and learn how these biological molecules can be identified in the lab.

Concepts

- Proteins
- Amino acids
- Peptide linkage
- Biuret test
- Xanthoproteic test

Background

Proteins represent the most diverse class of biological compounds within cells. It is estimated that a single bacteria cell contains more than 3,000 different proteins. The word protein is derived from the Greek word "proteios," meaning first or primary. Proteins are of primary importance in terms of both their occurrence within cells and their function in cell activities. The functions of proteins are at the center of life itself—proteins catalyze our metabolic reactions, carry oxygen to our body tissues, protect the body from infection, and maintain cell and tissue structure.

Proteins are composed of amino acid molecules joined together in chain-like fashion via *peptide linkages*. Amino acids are thus often referred to as the "building blocks" of protein structure. The number of amino acids in a single protein can vary from around 50 amino acid residues in insulin to more than 500 in hemoglobin and more than 5,000 in some viruses. When fewer than 50 amino acids are joined together, the resulting compounds are called polypeptides.

All amino acids have two structural groups in common—they contain a carboxylic acid group ($-COOH$) on one end and an amine group ($-NH_2$) on the other end. Peptide linkages are created when the carboxyl group of one amino acid reacts with the amino group of the next amino acid in the sequence. As each amino acid is added to the growing polypeptide chain, a molecule of water is formed as a byproduct, as shown in Figure 1.

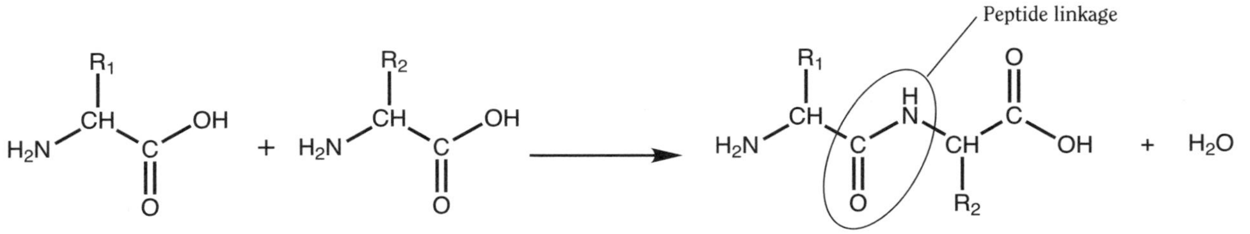

At neutral pH, almost all of the amino acids exist in the form of their neutral "zwitterions."

$$H_3\overset{+}{N}-CHR-\overset{-}{CO_2}$$

In order for amino acids to react with one another to form peptide linkages, the carboxyl groups must be activated. This is achieved biochemically through the use of transfer-RNA molecules that bond to the carboxyl group and make it more reactive.

Figure 1. Formation of a peptide linkage.

All proteins are made from about 20 different, naturally occurring amino acids, which can be arranged in an almost infinite number of ways. The primary structure of a protein is determined by the number and identity of amino acids and the order in which they are joined together. Higher levels of protein structure (called secondary, tertiary, and quaternary structure) result as the polypeptide chains form ribbons, sheets, and coils that then fold in on themselves to form more stable three-dimensional arrangements.

In addition to these reactive amine and carboxylic acid functional groups, amino acids contain a third group of atoms, called the side chain (shown as "R" groups in Figure 1). Although not involved in peptide bond formation, the side chains may contain other functional groups that influence both the structure and function of proteins. Hydrophobic amino acids contain nonpolar ("water-fearing") side chains, such as large hydrocarbon groups. Protein molecules often fold in on themselves so that the hydrophobic amino acids are tucked away in the interior of the structure. This reduces unfavorable contact between the nonpolar side chains and polar water molecules within cells. Amino acids are classified as hydrophilic if they contain polar ("water-loving") side chains that are able to form strong hydrogen bonds. Hydrophilic amino acids are often found at the "active" sites in enzymes, where they bind to small molecules and catalyze chemical reactions. Finally, ionic amino acids contain extra acidic and basic groups in their side chains; at physiological pH these side chains exist in charged, ionic forms. Oppositely charged side chains form so-called "salt bridges" that stabilize the three-dimensional structure of proteins.

Experiment Overview

The purpose of this experiment is to identify proteins and amino acids using a series of classification tests. Proteins will be identified using a simple color test based on the reaction of their polypeptide backbones with copper ions in basic solution. Compounds containing two or more peptide linkages react with copper sulfate to form a purple complex. This is called the *biuret test*. The purple product is due to coordination of peptide nitrogen atoms with copper ions. The amount of product that is formed and thus the intensity of the purple color depend on the nature of the protein and how much protein is present.

Specific amino acid residues in proteins will be identified using chemical tests based on reactions of their different side chains. The amino acid groups that will be identified in this lab are the aromatic side chain in tyrosine, the basic side chain in arginine, and the sulfur-containing side chain in cysteine. The aromatic amino acid tyrosine is identified by means of the *xanthoproteic test* (Greek for "yellow protein"). Reaction of tyrosine with nitric acid results in nitration of the aromatic ring to give a yellow product. Arginine is identified by means of the *Sakaguchi test,* which involves reaction with α-naphthol and sodium hypochlorite to give a red solution. Finally, the sulfur-containing amino acid cysteine will be identified using the *nitroprusside test,* which involves reaction with sodium nitroprusside (also called sodium nitroferricyanide) to give a purple or brown product. The structures of these three key amino acids are shown in Figure 2.

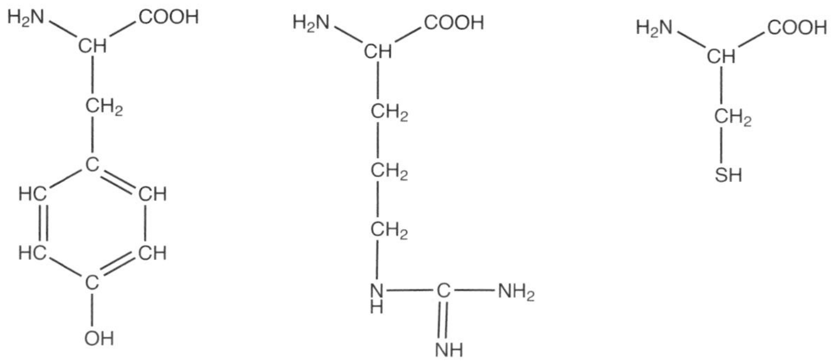

Figure 2. Structures of tyrosine, arginine, and cysteine.

Teacher Notes

Pre-Lab Questions

The popular low-calorie sweetener NutraSweet®
(aspartame) is prepared from the amino acids
phenylalanine and aspartic acid. The structure of
aspartame is shown to the right.

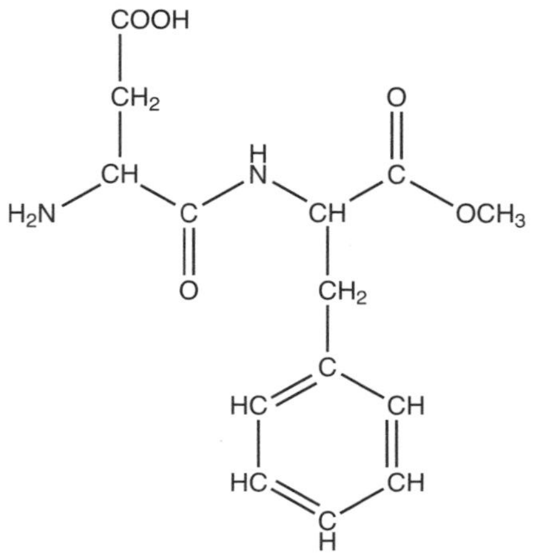

1. Circle and label the following groups in the
   structure of aspartame: the peptide linkage, the
   "terminal" amino group, and the hydrophobic
   amino acid side chain.

2. Which amino acid side chain would be expected
   to participate in hydrogen bonding?

3. Would you expect aspartame to give a positive
   biuret test? Explain.

Materials

Protein Solutions

    Albumin, 2%, 4 mL

    Casein, 2%, 4 mL

    Gelatin, 2%, 4 mL

Amino Acid Solutions

    Arginine, 1%, 4 mL

    Cysteine, 1%, 4 mL

    Tyrosine, 1%, 4 mL

Testing Solutions

    Biuret test solution, 7 mL

    α-Naphthol, 0.1% in ethyl alcohol, 2 mL

    Nitric acid, HNO_3, 3 M, 7 mL

    Sodium hydroxide, NaOH, 3 M, 10 mL

    Sodium hypochlorite (bleach), NaOCl, 5%, 30 mL

    Sodium nitroferricyanide, $Na_2Fe(CN)_5NO\cdot H_2O$, 2%, 5 mL

Beakers, 400- and 600-mL

Hot plate

Pipets, Beral-type, graduated, 12

Test tubes, small, 7

Test tube clamp

Test tube rack

Wash bottle

Water, distilled or deionized

Safety Precautions

*Biuret test solution contains copper sulfate, which is moderately toxic by ingestion, and
sodium hydroxide, which is corrosive to eye and body tissue. α-Naphthol is slightly toxic
by ingestion, inhalation, and skin absorption and is a body tissue irritant. The solution
contains ethyl alcohol and is a flammable liquid. Avoid contact with flames or other
sources of ignition. Nitric acid, sodium hydroxide, and sodium hypochlorite solutions are
corrosive liquids and can cause skin burns. Sodium nitroferricyanide is highly toxic by
ingestion and inhalation. Do not allow this solution to come in contact with acids. Do not
heat the solution. Dispense and use sodium nitroferricyanide in the hood or in a well-
ventilated lab only. Avoid exposure of all chemicals to eyes and skin. Wear chemical splash
goggles, chemical-resistant gloves, and a chemical-resistant apron.*

*Aspartame is a dipeptide
formed from the amino
acids aspartic acid and
phenylalanine. The
artificial sweetener is a
methyl ester derivative
of the parent dipeptide—
notice the —CO_2CH_3
group at the end of the
structure.*

Procedure

Prepare a boiling water bath for use in Part B. Fill a 400-mL beaker half-full with water and add a boiling stone. Heat the water bath to boiling on a hot plate.

Part A. Biuret Test

1. Label a set of seven test tubes 1–7.

2. Use a graduated Beral-type pipet to add 1 mL of each solution to be tested to the appropriate test tube, as follows:

Test tube	1	2	3	4	5	6	7
Solution	Water	Albumin	Casein	Gelatin	Arginine	Cysteine	Tyrosine

3. To each test tube add 1 mL of biuret test solution.

4. Observe the color and appearance of each solution and record the results in the Data Table.

5. Pour 25 mL of 5% sodium hypochlorite solution into a 600-mL beaker to be used for waste disposal. Set the beaker aside for use in steps 6, 11, 15, and 19.

6. Rinse the contents of the test tubes with a large amount of water into the waste disposal beaker. Wash the test tubes and rinse well with distilled water. Relabel them 1–7, if necessary, for use in the next test.

Part B. Xanthoproteic Test

7. Repeat step 2 to prepare a set of protein and amino acid samples to be tested.

8. To each test tube add 1 mL of 3 M nitric acid.

9. Place the test tubes in the boiling water bath for 3–5 minutes.

10. Use a test tube clamp to remove the test tubes from the boiling water bath. Allow the solutions to cool and then record your observations of their color and appearance in the Data Table.

11. Rinse the contents of the test tubes with a large amount of water into the waste disposal beaker. Wash the test tubes and rinse well with distilled water. Relabel them 1–7, if necessary, for use in the next test.

Part C. Sakaguchi Test

12. Repeat step 2 to prepare a set of protein and amino acid samples to be tested.

13. To each test tube add 3 drops of 3 M sodium hydroxide, followed by 5 drops of α-naphthol solution.

14. Add 10 drops of sodium hypochlorite to each test tube and record the color and appearance of each solution in the Data Table.

Teacher Notes

15. Rinse the contents of the test tubes with a large amount of water into the waste disposal beaker. Wash the test tubes and rinse well with distilled water. Relabel them 1–7, if necessary, for use in the next test.

Part D. Nitroprusside Test

16. Repeat step 2 to prepare a set of protein and amino acid samples to be tested.

17. To each test tube add 20 drops of 3 M sodium hydroxide, followed by 10 drops of sodium nitroferricyanide solution.

18. Record the color and appearance of each solution in the Data Table.

19. Rinse the contents of the test tubes with a large amount of water into the waste disposal beaker. Wash and rinse the test tubes.

20. Consult your instructor regarding proper disposal of the solution in the waste beaker.

Name: _____

Class/Lab Period: _____

Identifying Proteins and Amino Acids

Data Table

Classification Test	Test Tube and Sample						
	1	**2**	**3**	**4**	**5**	**6**	**7**
	Water	Albumin	Casein	Gelatin	Arginine	Cysteine	Tyrosine
Biuret Test							
Xanthoproteic Test							
Sakaguchi Test							
Nitroprusside Test							

Post-Lab Questions *(Use a separate sheet of paper to answer the following questions.)*

1. Which samples gave positive results and which gave negative results in the biuret test? Were there any differences in the color and intensity of the positive test results? How general is the biuret test for detecting proteins of different types?

2. Which amino acids are identified by means of the xanthoproteic test? Which protein samples gave positive xanthoproteic test results? Comment on the composition of the protein samples based on the results of this test.

3. Which amino acids are identified by means of the Sakaguchi test? Which protein samples gave positive Sakaguchi test results? Comment on the composition of the protein samples based on the results of this test.

4. Which amino acids are identified by means of the nitroprusside test? Which protein samples gave positive nitroprusside test results? Comment on the composition of the protein samples based on the results of this test.

Teacher Notes

Teacher's Notes
Identifying Proteins and Amino Acids

Master Materials List *(for a class of 30 students working in pairs)*

Protein Solutions*
  Albumin, 2%, 100 mL
  Casein, 2%, 100 mL
  Gelatin, 2%, 100 mL
Amino Acid Solutions*
  Arginine, 1%, 100 mL
  Cysteine, 1%, 100 mL
  Tyrosine, 1%, 100 mL
Testing Solutions
  Biuret test solution, 150 mL†
  α-Naphthol, 0.1% in ethyl alcohol, 50 mL
  Nitric acid, HNO_3, 3 M, 150 mL
  Sodium hydroxide, NaOH, 3 M, 200 mL
  Sodium hypochlorite (bleach), NaOCl, 5%, 500 mL
  Sodium nitroferricyanide, $Na_2Fe(CN)_5NO{\cdot}H_2O$, 2%, 100 mL

Beakers, 400- and 600-mL, 15 each
Pipets, Beral-type, graduated, 150–180
Hot plates, 2–3
Test tubes, 13 × 100 mm, 105
Test tube clamps, 15
Test tube racks, 15
Wash bottles, 15
Water, distilled or deionized

*Protein and amino acid solutions should be freshly prepared. See the *Preparation of Solutions* section.

†We recommend the purchase of ready-made Biuret test solution (Flinn Catalog No. B0051).

Preparation of Solutions *(for a class of 30 students working in pairs)*

For best results, prepare the protein and amino acid solutions no more than one week prior to lab. Cap the bottles and shake gently to dissolve. Vigorous shaking may cause foaming and denature the proteins.

Albumin, 2%: Dissolve 2 g of albumin in 100 mL of distilled or deionized water.

Casein, 2%: Casein is insoluble in water, but soluble in dilute base. Add 2 mL of 3 M NaOH solution to 98 mL of water, followed by 2 g of casein. Shake gently to dissolve.

Gelatin, 2%: Dissolve 2 g of gelatin in 100 mL of distilled or deionized water.

Arginine, 1%: Dissolve 1 g of arginine in 100 mL of distilled or deionized water.

Cysteine, 1%: Dissolve 1 g of cysteine in 100 mL of distilled or deionized water.

Tyrosine, 1%: Dissolve 1 g of tyrosine in 100 mL of distilled or deionized water.

α-Naphthol, 0.1%: Dissolve 0.05 g of α-napththol in 50 mL of ethyl alcohol (95%).

Nitric Acid, 3 M: Carefully add 29 mL of concentrated nitric acid (15.8 M) to 100 mL of distilled or deionized water. Stir to mix, then dilute to 150 mL.

Sodium Hydroxide, 3 M: Cool 100 mL of distilled or deionized water in an ice-water bath. Add 24 g of sodium hydroxide pellets and stir to dissolve. Bring to room temperature and dilute to 200 mL with water.

Sodium Nitroferricyanide, 2%: Dissolve 2 g of sodium nitroferricyanide dihydrate $(NaFe(CN)_5NO \cdot 2H_2O)$ in 100 mL of distilled or deionized water.

Safety Precautions

Biuret solution contains copper sulfate, which is moderately toxic by ingestion, and sodium hydroxide, which is corrosive to eye and body tissue. α-Naphthol solution contains ethyl alcohol and is a flammable liquid. Avoid contact with flames or other sources of ignition. Nitric acid, sodium hydroxide, and sodium hypochlorite solutions are corrosive liquids and can cause skin burns. Sodium nitroferricyanide solution is highly toxic by ingestion and inhalation. Do not allow this solution to come in contact with acids. Do not heat the solution. Dispense and use sodium nitroferricyanide in the hood or in a well-ventilated lab only. Avoid exposure of all chemicals to eyes and skin. Wear chemical splash goggles, chemical-resistant gloves, and a chemical-resistant apron. Please consult current Material Safety Data Sheets for further information on the safe use, handling, and disposal of these laboratory chemicals. Encourage students to wash their hands thoroughly before leaving the laboratory.

Disposal

Consult your current *Flinn Scientific Catalog/Reference Manual* for both general guidelines and specific methods governing the disposal of laboratory waste. All protein and amino acid samples may be flushed down the drain with excess water according to Flinn Suggested Disposal Method #26b. The waste solutions collected in the waste disposal beakers can be flushed down the drain with 20-fold excess water according to Flinn Suggested Disposal Method #26b. Excess sodium nitroferricyanide solution can be disposed of according to Flinn Suggested Disposal Method #14. Excess biuret test solution, sodium hydroxide, and sodium hypochlorite can be disposed of according to Flinn Suggested Disposal Method #10.

Lab Hints

- The experimental work for this lab can reasonably be completed in one 50-minute lab period.

- The classification tests can be performed in any order. To avoid congestion at the materials bench, consider staggering the starting points for different student groups. Set up separate stations for the four different tests in different locations and have students rotate among the stations to complete the lab activity. Another suggestion: Dispense beforehand smaller amounts of all of the solutions needed for each lab table or bench to use separately to prevent possible contamination of protein and amino acid samples.

- The name of the biuret test is frequently confusing to students and teachers alike. The test solution consists of copper sulfate in basic solution. The name of the test actually derives from the name of the reagent chemical, called biuret, that gives a characteristic positive control test with the test solution. Biuret is a derivative of urea that contains two amide ($-CONH_2$) groups in its structure and thus forms a purple coordination complex with copper ions. The biuret test solution itself does not contain any biuret!

Teacher Notes

- The xanthoproteic test was the subject of a famous public "Christmas" lecture by Michael Faraday, who demonstrated the reaction of silk and feathers with concentrated nitric acid. Consider performing this demonstration in class before the lab.

- The xanthoproteic test is normally performed with boiling, concentrated nitric acid. We have significantly improved the safety of this test by reducing the concentration of nitric acid to 3 M. The results are similar, except the color change to yellow is not always accompanied by a precipitate.

- Sodium nitroferricyanide (sodium nitroprusside) is highly toxic by ingestion and inhalation. Do not allow sodium nitroferricyanide solution to come in contact with strong acids. Do not heat the solution. Sodium nitroprusside is a pharmaceutical drug used in the treatment of high blood pressure and congestive heart disease. The nitroprusside test for cysteine is also a clinical test used in medical technology laboratories to test for excessive amounts of cysteine in urine, which can be a symptom of disease.

Teaching Tips

- In order to make the most efficient use of lab time, the ninhydrin reaction, a popular protein and amino acid test, has not been included. Ninhydrin is used to identify both amino acids and proteins and gives a positive, purple color test result with ALL of the samples included in this study. The ninhydrin test is most widely used to identify "spots" in the separation of amino acids by chromatography. See "Amino Acid Fingerprints" in the Demonstrations section of this Flinn ChemTopic™ Labs manual for an interesting application of the ninhydrin reaction in detecting amino acids in fingerprints.

Answers to Pre-Lab Questions *(Student answers will vary.)*

1. Circle and label the following groups in the structure of aspartame: the peptide linkage, the "terminal" amino group, and the hydrophobic amino acid side chain.

2. Which amino acid side chain would be expected to participate in hydrogen bonding?

See the following website for a complete description of Faraday's Christmas lectures and modern variations of his demonstrations.

www.woodrow.org/ teachers/chemistry/ institutes/faraday

3. Would you expect aspartame to give a positive biuret test? Explain.

Aspartame will not give a positive biuret test. The biuret reaction requires at least two peptide linkages in the reacting molecule.

Sample Data

Student data will vary.

Data Table

Classification Test	Test Tube and Sample						
	1	**2**	**3**	**4**	**5**	**6**	**7**
	Water	**Albumin**	**Casein**	**Gelatin**	**Arginine**	**Cysteine**	**Tyrosine**
Biuret Test	clear, pale blue solution	cloudy, purple solution	cloudy, purple solution	cloudy, lavender solution	clear, pale blue solution	clear, pale yellow solution	clear, pale blue solution
Xanthoproteic Test	clear, colorless solution	cloudy, yellow solution	cloudy solution, yellow precipitate	clear, colorless solution	clear, colorless solution	clear, colorless solution	cloudy solution, yellow precipitate
Sakaguchi Test	pale yellow solution	red solution	dark red solution	dark red solution	red solution	yellow-brown solution	yellow solution
Nitroprusside Test	yellow solution	brown solution	slightly brown solution	yellow solution	yellow solution	purple solution	yellow solution

The biuret test may be used for the quantitative determination of the amount of protein in solution. Prepare standard solutions containing 0.5, 1.0, 1.5, and 2% protein. Compare the intensity of the resulting purple color obtained with the biuret testing solution.

Answers to Post-Lab Questions *(Student answers will vary.)*

1. Which samples gave positive results and which gave negative results in the biuret test? Were there any differences in the color and intensity of the positive test results? How general is the biuret test for detecting proteins of different types?

   All of the proteins gave positive test results (lavender or purple solutions) with the biuret test. All of the amino acids examined in this study (arginine, cysteine, and tyrosine) gave negative results in the biuret test. Cysteine gave an anomalous, negative result. Most negative tests were the same pale blue as the distilled water blank. Cysteine gave a yellow solution, possibly due to some other reaction with copper ion. The different proteins did give qualitatively different results; the purple color was most intense for albumin and least intense for gelatin. The biuret test is a reliable general method for identifying proteins, since all of the proteins tested gave positive results. In order to be used for quantitative analysis of the amount of protein in a sample, however, the biuret test should first be "standardized" using known amounts of the protein to be analyzed. The intensity of the purple color that develops can then be compared against standard solutions to determine how much protein is present.

2. Which amino acids are identified by means of the xanthoproteic test? Which protein samples gave positive xanthoproteic test results? Comment on the composition of the protein samples based on the results of this test.

   The xanthoproteic test identifies the aromatic amino acid tyrosine. The proteins albumin and casein also gave positive results with this test. The fact that gelatin gave negative test results suggests that gelatin does not contain noticeable amounts of the amino acid tyrosine.

3. Which amino acids are identified by means of the Sakaguchi test? Which protein samples gave positive Sakaguchi test results? Comment on the composition of the protein samples based on the results of this test.

   The Sakaguchi test identifies the basic amino acid arginine. All of the proteins tested (albumin, casein, and gelatin) also gave positive results with this test. This suggests that all of these proteins contain the amino acid arginine in amounts great enough to give a positive Sakaguchi test.

4. Which amino acids are identified by means of the nitroprusside test? Which protein samples gave positive nitroprusside test results? Comment on the composition of the protein samples based on the results of this test.

   The nitroprusside test identifies the sulfur-containing amino acid cysteine. The proteins albumin and casein also gave positive results with this test, as evidenced by the color change from yellow to brown. The fact that gelatin gave negative test results suggests that gelatin does not contain noticeable amounts of the amino acid cysteine.

Physical Properties of Proteins

Introduction

The effects of acids and bases, inorganic salts, organic solvents, and temperature on the physical properties of proteins can help us understand the structures of proteins and how they fulfill their vital biological functions.

Concepts

- Protein folding
- Denaturation
- Native structure
- Salting out

Background

Structure often relates to function—nowhere is this relationship more evident than in the description of the structures, physical properties, and biological roles of proteins. The structure of hemoglobin allows it to bind to oxygen and deliver oxygen to body tissues. The structure of a specific antibody protein allows it to recognize, bind, and destroy a potentially harmful foreign substance. The structure of collagen makes skin both elastic and strong.

The biological activity of a protein depends on its three-dimensional shape. All proteins have a common structural "backbone"—amino acid building blocks joined by chemical bonds called peptide linkages. There are more than 20 different, naturally occurring amino acids that differ in the types of atoms attached to the polypeptide backbone. The amino acid side chains—which can be large or small, polar or nonpolar, acidic or basic, positively or negatively charged—interact through a variety of forces. These forces include hydrogen bonding involving side chain –OH groups, dipole interactions among polar amino acids, ionic "salt bridges" between positively and negatively charged side chains, and hydrophobic effects that stabilize large, nonpolar side chains. All of these forces cause protein chains to twist and fold back on themselves into globular or spherical shapes. The forces that maintain the structure of proteins are illustrated schematically in Figure 1.

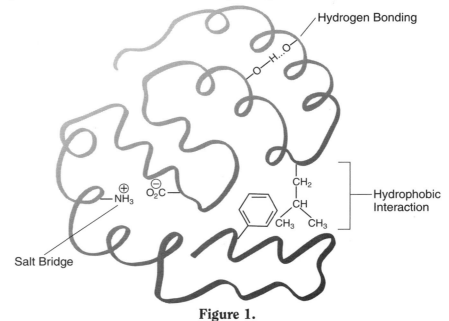

Figure 1.

Ribbon diagrams, similar to that shown in Figure 1, have become one of the most popular ways to illustrate protein structures. Stripping away the details of protein structure in this way clearly reveals the elements of secondary structure in proteins—alpha-helices, beta-pleated sheets, and random coils.

Protein folding is the name given to the process by which proteins naturally coil around or fold in on themselves in order to form a stable three-dimensional structure. Since every protein has a unique sequence of amino acids, every protein also has a unique shape—called its *native structure*—that makes the protein both stable and functional.

Denaturation

Any factor that disrupts the native structure of a protein will destroy its function. Destruction of the three-dimensional shape of a protein by physical or chemical means is called *denaturation*. Proteins become denatured by any action which breaks hydrogen bonds, destroys salt bridges, or interferes with hydrophobic interactions. Denaturation causes protein molecules to clump together and precipitate out of solution; the resulting loss of biological activity is generally irreversible. Heating, freezing, and agitation are physical processes that result in protein denaturation. Chemical agents that cause protein denaturation include strong acids and bases, organic solvents, and heavy metal salts.

Most proteins are denatured by temperatures above 50 °C (normal body temperature is 37 °C). Cooking an egg provides an everyday example of the changes that occur when a protein solution—the egg white—is heated. Heat supplies excess energy that disrupts the major forces holding a protein together. Strong acids or bases affect the number of charges on amino acid side chains and interfere with ionic "salt bridge" formation in proteins. Proteins have an optimal pH range where they are most soluble and most active. Small pH changes around the optimum pH may reduce the solubility of a protein, but are usually reversible. High concentrations of strong acid and strong base, on the other hand, precipitate proteins and lead to total loss of protein structure and function—irreversible denaturation. Proteins can also be denatured by the addition of polar organic solvents, such as alcohols and acetone, that interfere with hydrogen bonding. Heavy metal salts containing Ag^+, Hg^{2+}, and Pb^{2+} ions are poisonous as well because they denature proteins.

High concentrations of inorganic salts, such as ammonium sulfate, are used to precipitate proteins without loss of protein activity. The solubility of a protein decreases as the concentration of ionic compounds increases and the protein eventually precipitates out. This process—called *salting out*—results from changes in hydrogen bonding between water molecules and the protein. Because salting out involves mild conditions, the process is generally reversible. Salting out is used as a means of isolating and purifying proteins.

Experiment Overview

The purpose of this experiment is to examine the effects of acids and bases, inorganic salts, organic solvents, and temperature on the physical properties of proteins.

Pre-Lab Questions

1. Define the term denaturation. What is the most common, visible change that indicates denaturation has occurred?

2. Isopropyl alcohol is sold in drugstores as "rubbing alcohol," a disinfectant. What effect might alcohols have on bacterial proteins?

Demonstrate the effect of strong acid on proteins using an egg white. Place a raw egg white in a Petri dish on an overhead projector. Slowly add concentrated hydrochloric acid or sulfuric acid dropwise to the egg white. The egg white will immediately discolor and coagulate. Adding strong base will not reverse the process.

Teacher Notes

Materials

Albumin, 2% solution, 22 mL

Ammonium sulfate solution,
   $(NH_4)_2SO_4$, saturated, 25 mL

Casein, 2% solution, 2 mL

Copper(II) sulfate solution, $CuSO_4$, 0.1 M, 4 mL

Gelatin, 2% solution, 2 mL

Hydrochloric acid, HCl, 3 M, 6 mL

Isopropyl alcohol, $(CH_3)_2CHOH$, 2 mL

Silver nitrate solution, $AgNO_3$, 0.1 M, 2 mL

Sodium hydroxide solution, NaOH, 3 M, 5 mL

Water, distilled or deionized

Wash bottle

Beakers, 50- and 250-mL

Beral-type pipets, graduated, 9

Erlenmeyer flask, 125-mL

Filter paper and funnel

Hot plate

Stirring rod

Test tubes, small, 3

Test tube, medium, 1

Test tube clamp

Test tube rack

Thermometer

Safety Precautions

Hydrochloric acid and sodium hydroxide solutions are corrosive liquids and can cause skin burns. Silver nitrate solution is a corrosive liquid and toxic by ingestion; it will stain skin and clothes. Isopropyl alcohol is a flammable organic solvent; do not use near flames or other sources of ignition. Ammonium sulfate and copper sulfate solutions are slightly toxic by ingestion. Avoid exposure of all chemicals to eyes and skin. Wear chemical splash goggles, chemical-resistant gloves, and a chemical-resistant apron. Wash hands thoroughly with soap and water before leaving the laboratory.

Procedure

Part A. Solubility and Protein Denaturation

1. Label three small test tubes 1–3.

2. Using a clean, graduated Beral-type pipet for each solution, add approximately 1 mL of albumin, casein, and gelatin to test tubes 1, 2, and 3, respectively. Record the initial appearance of each solution in Data Table A.

3. Add 2 drops of 3 M hydrochloric acid to each test tube 1–3. Gently swirl each tube to mix the contents, then record the appearance of the solutions in Data Table A.

4. Add 5 more drops of 3 M hydrochloric acid to each test tube 1–3. Swirl each sample mixture and record its appearance in Data Table A.

5. Add 10 more drops of 3 M hydrochloric acid to each test tube 1–3. Swirl each sample mixture and record its appearance in Data Table A.

6. Add 10 more drops of 3 M hydrochloric acid to each test tube 1–3. Swirl each sample mixture and record its appearance in Data Table A.

7. Wash the contents of each test tube down the drain with excess water and rinse the test tubes twice with distilled water from a wash bottle. Relabel the test tubes 1–3, if necessary.

8. Using the appropriate graduated Beral-type pipet for each solution, add approximately 1 mL of albumin, casein, and gelatin to test tubes 1, 2, and 3, respectively.

9. Add 5 drops of 3 M sodium hydroxide to each test tube 1–3. Gently swirl each tube to mix the contents and record the appearance of the solutions in Data Table A.

10. Add 10 more drops of 3 M sodium hydoxide to each test tube 1–3. Swirl each sample mixture and record its appearance in Data Table A.

11. Wash the test tube contents down the drain with excess water and rinse the test tubes twice with distilled water. Relabel the test tubes 1–3, if necessary.

12. Add 1 mL of 2% albumin solution to each tube.

13. Using a clean, graduated Beral-type pipet for each reagent, add 2 mL of 0.1 M copper(II) sulfate to test tube 1, 2 mL of 0.1 M silver nitrate to test tube 2, and 2 mL of isopropyl alcohol to test tube 3. Gently swirl each tube to mix the contents, then record their appearance in Data Table A.

Part B. "Salting Out" with Ammonium Sulfate

14. Add 10 mL of 2% albumin to a 50-mL beaker, followed by approximately 25 mL of saturated ammonium sulfate solution. Stir the mixture thoroughly using a glass stirring rod. Describe the appearance of the mixture in Data Table B.

15. Set up a gravity filtration apparatus (see Figure 2) and filter the mixture through a piece of wetted filter paper. Collect the liquid (filtrate) in a clean Erlenmeyer flask.

16. Label three small test tubes 1–3.

    * Add 2 mL of the original 2% albumin solution to test tube 1.
    * Add 2 mL of the filtrate from step 15 to test tube 2.
    * Remove a small portion of the precipitate from the funnel with the tip of a spatula, and dissolve the wet solid in 2 mL of distilled water in test tube 3.

17. To each test tube 1–3, add 10 drops of 3 M sodium hydroxide followed by 5 drops of 0.1 M copper(II) sulfate solution. Compare the appearance of the three solutions and record the observations in Data Table B.

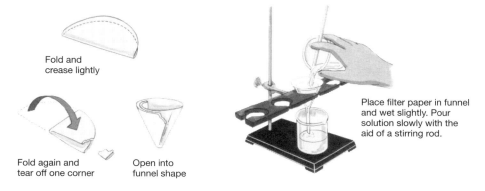

Fold and crease lightly

Fold again and tear off one corner

Open into funnel shape

Place filter paper in funnel and wet slightly. Pour solution slowly with the aid of a stirring rod.

Figure 2. Gravity filtration setup.

The reaction of the solutions in step 16 with $CuSO_4$ and NaOH is an example of the biuret test. See the previous experiment, "Identifying Proteins and Amino Acids," in this lab book for a description and explanation of the biuret test.

Part C. Effect of Heat

18. Prepare a hot water bath: Fill a 250-mL beaker half-full with water and heat it on a hot plate at the lowest setting. Place a thermometer in the water bath to record the temperature of the bath.

19. To a medium size test tube, add 5 mL of 2% albumin solution.

20. When the temperature of the hot water bath is 35–40 °C, place the test tube in the bath. Record the initial temperature of the water bath in Data Table C. Adjust the heat setting on the hot plate to a medium-high range to slowly heat the protein solution.

21. Holding the test tube with a test tube clamp, gently swirl the protein solution and observe its appearance. Note the temperature of the bath when the first signs of protein precipitation are observed. Record the temperature and make observatons in Data Table C.

22. Continue heating the protein solution. Record the temperature of the hot water bath and make observations of the protein solution when it first appears milky white (opaque).

23. When the temperature of the hot water bath reaches 85–90 °C, remove the test tube. Record the final appearance of the protein sample in Data Table C.

24. Dispose of the test tube contents as instructed by your teacher.

Different proteins show different heat sensitivities. If time permits, have students test the effect of heat on the three different proteins used in this experiment.

Name: _____

Class/Lab Period: _____

Physical Properties of Proteins

Data Table A. *Solubility and Protein Denaturation*

Effect of Strong Acid and Base				
Test Tube		**1**	**2**	**3**
Protein		Albumin	Casein	Gelatin
Initial Appearance				
Effect of HCl	2 drops			
	5 drops			
	10 drops			
	10 drops			
Effect of NaOH	5 drops			
	10 drops			

Effect of Inorganic and Organic Additives			
Test Tube	**1**	**2**	**3**
Additive	$CuSO_4$	$AgNO_3$	Isopropyl Alcohol
Results			

Data Table B. *"Salting Out" with Ammonium Sulfate*

	Observations
Effect of Ammonium Sulfate	
Test Tube 1 **(Albumin + $CuSO_4$)**	
Test Tube 2 **(Filtrate + $CuSO_4$)**	
Test Tube 3 **(Redissolved solid + $CuSO_4$)**	

Teacher Notes

Data Table C. *Effect of Heat*

	Temperature	Additional Observations
Initial temperature (water bath)		
First signs of precipitate appeared		
Solution appeared milky white		
Final observations		

Post-Lab Questions *Use a separate sheet of paper to answer the following questions.*

1. Compare and contrast the effect of strong acid (HCl) on albumin, casein, and gelatin. Which protein was most sensitive to the action of strong acid? Least sensitive?

2. Do strong acids and strong bases have similar effects on protein solubility and denaturation? Explain.

3. Which metal salts ($CuSO_4$ and $AgNO_3$) caused albumin denaturation? How does this observation relate to the toxicity of silver salts versus copper salts?

4. You have just been to the doctor's office to receive an inoculation. Before administering the injection, the doctor wipes the area with an alcohol swab. Do the results for the effect of alcohol on albumin denaturation support the use of isopropyl alcohol as a disinfectant? Explain.

5. The reaction of $CuSO_4$ with proteins in strong base is used as a color test to identify proteins. What do the results obtained in Part B for the reaction of $CuSO_4$ with albumin (test tube 1) and the filtrate (test tube 2) tell you about the effectiveness of the "salting out" procedure with ammonium sulfate?

6. Is the denaturation of albumin by ammonium sulfate reversible or irreversible? Explain on the basis of your observations for the reaction of $CuSO_4$ with albumin (test tube 1) and the redissolved precipitate (test tube 3), respectively, in Part B.

7. Based on the results of Part C, explain why heat is an effective form of sterilization for biological materials and equipment.

Teacher's Notes
Physical Properties of Proteins

Master Materials List *(for a class of 30 students working in pairs)*

Albumin, 8 g*

Ammonium sulfate solution,
  $(NH_4)_2SO_4$, saturated, 400 mL

Casein, 1 g*

Copper(II) sulfate solution,
  $CuSO_4$, 0.1 M, 100 mL

Gelatin, 1 g*

Hydrochloric acid, HCl, 3 M, 100 mL

Isopropyl alcohol, $(CH_3)_2CHOH$, 50 mL

Silver nitrate solution, $AgNO_3$, 0.1 M, 50 mL

Sodium hydroxide solution, NaOH, 3 M, 100 mL

Water, distilled or deionized

Beakers, 50- and 250-mL, 15

Erlenmeyer flasks, 125-mL, 15

Funnels and filter paper, 15

Hot plates, 3–5

Pipets, Beral-type, graduated, 135

Stirring rods, 15

Test-tubes, 13 × 100 mm, 45

Test-tube, 16 × 125 mm, 15

Test tube clamps, 15

Test tube racks, 15

Thermometers, 15

Wash bottles, 15

*Protein solutions must be freshly prepared. See the *Preparation of Solutions* section.

Preparation of Solutions *(for a class of 30 students working in pairs)*

For best results, prepare the protein solutions no more than one week prior to lab. Cap the bottles and shake gently to dissolve. More extreme shaking or agitation of protein solutions may cause the proteins to denature and precipitate out of solution.

Albumin, 2%: Dissolve 8 g of albumin in 250 mL of distilled or deionized water, then dilute to 400 mL.

Ammonium Sulfate, Saturated: Dissolve 280 g of ammonium sulfate in 400 mL of distilled or deionized water. Stir the mixture for at least one hour to dissolve as much of the solid as possible. Decant or filter the solution and use the clear filtrate.

Casein, 2%: Casein is insoluble in water, soluble in dilute base. Add 1 mL of 3 M NaOH to 49 mL of distilled or deionized water, followed by 1 g of casein.

Copper(II) Sulfate, 0.1 M: Add 2.5 g of cupric sulfate pentahydrate to 50 mL of distilled or deionized water. Stir to dissolve, then dilute to 100 mL.

Gelatin, 2%: Dissolve 1 g of gelatin in 50 mL of distilled or deionized water. Shake gently to dissolve.

Hydrochloric Acid, 3 M: Place about 50 mL of distilled or deionized water in a flask and add 25 mL of concentrated (12 M) hydrochloric acid. Stir to mix, then dilute to 100 mL. *Note:* Always add acid to water.

The concentration of ammonium sulfate required to "salt-out" different proteins depends on the nature of the protein. In this experiment, albumin requires a very high concentration of ammonium sulfate to precipitate out of solution.

Silver Nitrate, 0.1 M: Add 0.85 g of silver nitrate to about 30 mL of distilled or deionized water. Stir to dissolve, then dilute to 50 mL.

Sodium Hydroxide, 3 M: Add 12 g of sodium hydroxide to about 60 mL of distilled or deionized water. Stir to dissolve, then allow to cool and dilute to 100 mL.

Safety Precautions

Hydrochloric acid and sodium hydroxide solutions are corrosive liquids and can cause skin burns. Silver nitrate solution is a corrosive liquid and toxic by ingestion; it will stain skin and clothes. Isopropyl alcohol is a flammable organic solvent; do not use near flames or other sources of ignition. Ammonium sulfate and copper sulfate solutions are slightly toxic by ingestion. Avoid exposure of all chemicals to eyes and skin. Wear chemical splash goggles, chemical-resistant gloves, and a chemical-resistant apron. Please consult current Material Safety Data Sheets for safety, handling, and disposal information.

Disposal

Consult your current *Flinn Scientific Catalog/Reference Manual* for both general guidelines and specific methods governing the disposal of laboratory chemicals. All protein solutions and test mixtures from Parts A, B, and C can be flushed down the drain with excess water according to Flinn Suggested Disposal Method #26b.

Lab Hints

- The experiments in this activity can be performed in any order. To avoid congestion, consider staggering the starting points for different student groups. The experimental work for this lab can reasonably be completed in one 50-minute lab period.

- Dispense beforehand smaller amounts of all of the solutions for each lab table or bench to use separately. Encourage students to label their Beral-type pipets to prevent contamination and waste.

- Albumin is the chief protein in egg white. It serves as a source of amino acids for the developing embryo. Casein is the principal protein in milk. It has a high concentration of phosphate groups attached to its amino acid residues and is also associated with the high calcium content in milk. Casein is readily precipitated from milk with dilute acid— it has its minimum solubility at a pH of 4.7. Gelatin is a mixture of proteins obtained by hydrolysis of collagen from animal skin, ligaments, and tendons. Because of the way it is prepared, gelatin consists of shorter molecular weight protein fragments that are relatively insensitive to denaturation by acids.

- The reaction of copper sulfate with proteins in the presence of sodium hydroxide (Part B) is called the biuret test and is a common color test to identify proteins. Copper ions coordinate with the amide functional groups in the polypeptide backbone to form highly colored, purple complex ions. For a more complete description of the biuret test, see the previous experiment in this book, "Identifying Proteins and Amino Acids."

- The effect of HCl on protein solubility and denaturation provides an excellent opportunity to reinforce safety rules concerning the corrosive nature of strong acids. Consider leading into this lab with a safety demonstration that shows how strong acid affects the protein in egg white. Call or write us at Flinn Scientific and we will fax you a free handout of our *SafetyFax* demonstration: Publication No. 801.00, "Acid in the Eye." Your students will never again forget to wear their safety goggles!

• An alternative way to study the heat denaturation of proteins is to set up several different temperature baths in the classroom and have students measure the time it takes for albumin to coagulate and precipitate at different temperatures. Three hot water baths, at 40, 60, and 80 °C, should be enough to gather interesting data for students to compare. This experiment can also be extended to examine the "denaturation temperature" of different proteins, which vary in their sensitivity to heat. Albumin is one of the most heat-sensitive proteins. This could lead to a discussion of heat-resistant proteins that are present in unusual bacteria, such as those that thrive in hot springs.

Teaching Tips

• The Internet is a wonderful resource to explore the structure and properties of proteins. The "Protein Data Bank" (http://www.rcsb.org/pdb) is a searchable database maintained by collaboration of academic and government scientists. It offers access to thousands of protein structures.

• Protein folding is the subject of intense current research. Consider holding a class seminar to discuss current theories of how and why proteins adopt the shapes they do. There is even evidence that mistakes in protein folding may give rise to a variety of diseases, including Alzheimer's disease and cystic fibrosis.

Answers to Pre-Lab Questions *(Student answers will vary.)*

1. Define the term denaturation. What is the most common, visible change that indicates denaturation has occurred?

   Denaturation refers to the loss of biological activity that occurs when the three-dimensional structure of a protein is disrupted due to physical conditions or chemical treatment. Denaturation is usually identified when the protein precipitates out of solution.

2. Isopropyl alcohol is sold in drugstores as "rubbing alcohol," a disinfectant. What effect might alcohols have on bacterial proteins?

   Alcohols disrupt hydrogen bonding in proteins and denature the proteins. When essential membrane proteins in bacteria are denatured, bacteria are killed because they lack proteins needed to stay alive.

Sample Data

Student data will vary.

Data Table A. *Solubility and Protein Denaturation*

Effect of Strong Acid and Base				
Test Tube		**1**	**2**	**3**
Protein		**Albumin**	**Casein**	**Gelatin**
Initial Appearance		cloudy solution	cloudy solution	cloudy solution
Effect of HCl	**2 drops**	clear solution	slight precipitate	clear solution
	5 drops	slight precipitate	clumpy white precipitate	clear solution
	10 drops	milky white mixture	large amount clumpy precipitate	clear solution
	10 drops	large amount white precipitate	large amount white precipitate	clear solution
Effect of NaOH	**5 drops**	clear solution	clear solution	clear solution
	10 drops	trace precipitate	clear solution	clear solution

Effect of Inorganic and Organic Additives		
Test Tube	**1**	**2**	**3**
Additive	$CuSO_4$	$AgNO_3$	**Isopropyl Alcohol**
Results	cloudy, blue solution	clumpy white precipitate	milky white (opaque) mixture white precipitate

Data Table B. *"Salting Out" with Ammonium Sulfate*

	Observations
Effect of Ammonium Sulfate	large amount of clumpy white precipitate settled out of solution
Test Tube 1 **(Albumin + $CuSO_4$)**	cloudy, purple solution
Test Tube 2 **(Filtrate + $CuSO_4$)**	clear, pale blue solution
Test Tube 3 **(Redissolved solid + $CuSO_4$)**	cloudy, purple solution

Data Table C. *Effect of Heat*

	Temperature	Additional Observations
Initial temperature (water bath)	35 °C	cloudy solution
First signs of precipitate appeared	47 °C	small amount of white precipitate floating in solution
Solution appeared milky white	70 °C	opaque, milky-white mixture
Final observations	85 °C	thick, gel-like mixture; clumpy white precipitate

Answers to Post-Lab Questions *(Student answers will vary.)*

1. Compare and contrast the effect of strong acid (HCl) on albumin, casein, and gelatin. Which protein was most sensitive to the action of strong acid? Least sensitive?

   Adding HCl to albumin and casein led to rapid coagulation and protein precipitation. Casein was the most sensitive to acid—only a few drops of HCl caused the casein to settle out of solution. Gelatin was the least affected by acid treatment; in fact, gelatin did not precipitate even after more than 25 drops of HCl had been added.

2. Do strong acids and strong bases have similar effects on protein solubility and denaturation? Explain.

   No, strong acid and strong base do not have the same effect on protein solubility. Protein denaturation was not observed even after 15 drops of 3 M NaOH were added. Actually, addition of base seemed to increase the solubility of proteins, since the initial cloudy protein solutions turned clear.

3. Which metal salts (CuSO$_4$ and AgNO$_3$) caused albumin denaturation? How does this observation relate to the toxicity of silver salts versus copper salts?

   AgNO$_3$ denatured albumin (a white precipitate was observed), whereas CuSO$_4$ did not affect the solubility. This may be related to the biological role of these metal ions. Although copper salts are slightly toxic, copper(II) ions are important enzyme cofactors and play an essential role in metabolism. Silver ion is highly toxic (although it does not pose as harmful an environmental hazard as mercury and lead). The toxicity of heavy metal salts is generally attributed to irreversible denaturation of proteins.

4. You have just been to the doctor's office to receive an inoculation. Before administering the injection, the doctor wipes the area with an alcohol swab. Do the results for the effect of alcohol on albumin denaturation support the use of isopropyl alcohol as a disinfectant? Explain.

   Alcohol is an effective disinfectant when applied to the skin. Alcohol denatures essential proteins in bacteria and kills the bacteria. In our experiment, adding isopropyl alcohol to albumin caused the protein to denature and precipitate out of solution.

5. The reaction of $CuSO_4$ with proteins in strong base is used as a color test to identify proteins. What do the results obtained in Part B for the reaction of $CuSO_4$ with albumin (test tube 1) and the filtrate (test tube 2) tell you about the effectiveness of the "salting out" procedure with ammonium sulfate?

   When copper sulfate was added to albumin (test tube 1) in the presence of sodium hydroxide, a purple color was observed. This serves as a positive color test to identify the protein. The filtrate (test tube 2) did not give a positive color test—it remained pale blue, the original color of copper sulfate by itself. This indicates that the filtrate did not contain any residual albumin, that in fact all of the protein was "salted out" by the addition of ammonium sulfate.

6. Is the denaturation of albumin by ammonium sulfate reversible or irreversible? Explain on the basis of your observations for the reaction of $CuSO_4$ with albumin (test tube 1) and the redissolved precipitate (test tube 3), respectively, in Part B.

   Denaturation of albumin by ammonium sulfate is reversible. This is demonstrated by two observations: the insoluble, denatured protein easily redissolved when water was added and the resulting protein solution (test tube 3) gave a positive test with copper sulfate and sodium hydroxide, just like the albumin control (test tube 1).

7. Based on the results of Part C, explain why heat is an effective form of sterilization for biological materials and equipment.

   Heat is an effective form of sterilization because it destroys essential proteins that bacteria and viruses might need to survive.

Properties of Lipids

Introduction

Fats and oils, waxes and cholesterol, steroid hormones and Vitamins A and D—all of these natural products belong to the diverse class of biological compounds called lipids. What are the properties of lipids? What role do lipids play in the chemistry of life?

Concepts

- Lipids
- Polar and nonpolar compounds
- Triglycerides
- Fats and oils
- Saturated vs. unsaturated
- Extraction

Background

Biological substances that are insoluble in water are classified as lipids. This characteristic physical property of lipids makes them quite different from other types of biological compounds—carbohydrates, proteins, and nucleic acids—that readily dissolve in water. Lipids do dissolve in nonpolar organic solvents, such as hexane, ether, and toluene, and are usually obtained from plant and animal tissues by extraction with an organic solvent.

The structures of lipids are extremely varied. Examples of lipids include:

- Fats and oils, such as butter and corn oil, that are familiar to us from nutrition.
- Phospholipids, also known as membrane lipids, that make up the cell membranes of all living organisms.
- Steroid hormones, such as estrogen and progesterone, that regulate cell activity.

The biological functions of lipids are as diverse as their structures. Fats and oils, for example, are used to store energy within cells and organisms. In addition to acting as energy sources, fats accumulate in adipose tissue that insulates and protects internal organs. Phospholipids are responsible for the "lipid bilayer" structures that form protective membranes around cells. The steroid hormones, which are synthesized in the body from cholesterol, act as chemical messengers, carrying signals from one part of the body to another.

Triglycerides

Fats and oils—referred to collectively as triglycerides—have the same basic structure. Triglycerides consist of a glycerol backbone and three attached fatty acid residues. Fatty acids are long-chain carboxylic acids, consisting of a long hydrocarbon "tail" (CH_3—CH_2—CH_2—CH_2—) with a carboxyl group (—COO) at one end. Fatty acids range in length from 10 to 20 carbon atoms and always contain an even number of carbon atoms. The two most common numbers of carbon atoms are 16 and 18. The hydrocarbon chains in fatty acids can be saturated or unsaturated. Saturated fatty acids contain only C—C single bonds in the hydrocarbon chain, while unsaturated fatty acids contain at least one C=C double bond. The presence of C=C double bonds reduces the number of hydrogen atoms in the hydrocarbon tail—these fatty acids are "unsaturated" with respect to the number of hydrogen atoms. Fatty acids that contain more than one double bond are called polyunsaturated.

Begin the discussion of lipids by asking students to bring in nutrition labels from their favorite food items. In addition to partially hydrogenated vegetable oils, the most common lipid in foods is lecithin, a membrane phospholipid isolated from egg yolk. Lecithin is added as an emulsifier to foods. Nutrition labels also provide a wealth of information concerning saturated and unsaturated fats.

Properties of Lipids

Fats are solids, obtained primarily from animal tissue, that contain a large proportion of saturated fatty acids. Oils are liquids, obtained primarily from plants, that contain a greater proportion of unsaturated fatty acids. This key structural difference has important consequences in nutrition—replacing saturated fats in the diet with polyunsaturated oils may help prevent heart disease. A close look at the nutritional label attached to any food item reveals not only the amount of "fat" in the food, but also the amount of saturated, monounsaturated, and polyunsaturated fats. The role of saturated versus unsaturated fats in nutrition is related to their structures. Unsaturated fatty acids have bends in their structures at the location of the C═C double bonds, and these bends make oils more fluid and less rigid than fats. Unsaturated fats may prevent the buildup of solid residues in arteries and veins. Figure 1 shows the structure of a triglyceride containing both saturated and unsaturated fatty acids.

$$CH_2-O-\overset{\overset{O}{\|}}{C}-CH_2-CH_2-CH_2-CH_2-CH_2-CH_2-CH_2-CH=CH-CH_2-CH=CH-CH_2-CH_2-CH_2-CH_2-CH_3$$

$$CH-O-\overset{\overset{O}{\|}}{C}-CH_2-CH_2-CH_2-CH_2-CH_2-CH_2-CH_2-CH_2-CH_2-CH_2-CH_2-CH_2-CH_2-CH_2-CH_3$$

$$CH_2-O-\overset{\overset{O}{\|}}{C}-CH_2-CH_2-CH_2-CH_2-CH_2-CH_2-CH_2-CH=CH-CH_2-CH_2-CH_2-CH_2-CH_2-CH_2-CH_2-CH_3$$

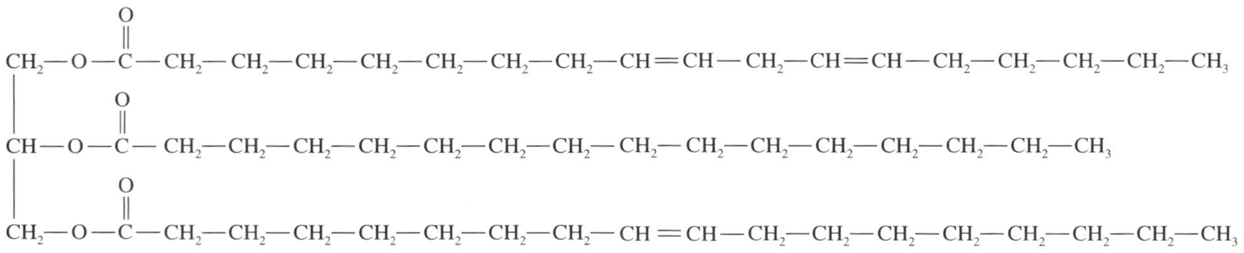

Figure 1.

Whether a triglyceride is a solid or liquid at room temperature also depends on the average number of carbon atoms. Coconut oil, for example, is considered one of the "unhealthy" oils—it contains more than 90% saturated fatty acids. However, because it has a high proportion of short-chain fatty acids (the average chain length is about 12), coconut oil is a liquid at room temperature.

There are two principal methods of obtaining the so-called seed oils (corn oil, olive oil, canola oil, etc.) from seeds. Unrefined oils are obtained by "squeezing" seeds under high pressure at elevated temperatures. Refined oils are obtained by extraction—the ground seeds are stirred with hexane or other petroleum solvents, which dissolve the oils. The resulting extracts are then heated to remove the solvent and the oils are subjected to further heat processing to improve their shelf life and stability.

Overview of the Experiments

The purpose of this experiment is to identify and classify lipids and examine their properties. In Part A, the solubility of lipids will be studied and compared with that of albumin, a protein isolated from eggs. The Sudan III test—a classic test for identifying lipids—will also be run. Sudan III is a special dye that is attracted to nonpolar compounds. It is used as a "fat stain" in botany and medicine to identify lipids in seeds and tissue samples, respectively.

In Part B, the presence of unsaturation in oils will be detected by mixing the oils with bromine water. Bromine (Br_2) combines with C═C double bonds to form "dibromides," in which the bromine atoms have added to the two carbon atoms in the double bond. Saturated compounds do not react with bromine. Disappearance of the orange-red color due to bromine serves as a positive test for unsaturation.

Teacher Notes

In Part C, the amount of "peanut oil" in peanuts will be determined by extracting peanuts with hexane. By measuring the mass of peanuts before and after solvent extraction, the amount of lipids can be determined and compared with the information provided on the nutritional label for peanuts.

Pre-Lab Questions

1. The C=C double bonds in unsaturated oils react with hydrogen to give "hydrogenated" compounds. What food items on your kitchen shelves have hydrogenated oils listed as a major ingredient? *Hint:* The snack food section is a good place to look.

2. What should happen to the consistency of an oil when it is hydrogenated? What are the advantages of this change in consistency in food processing? What are the disadvantages?

3. The fat content in milk can be measured by extraction with hexane. A 50.0-g sample of whole milk was mixed with hexane and gently stirred. Two liquid layers were obtained—a lower, "milky" layer and an upper layer of hexane. The layers were separated and the mass of the milky layer was found to be 47.8 g. Calculate the percentage of fat in whole milk.

Materials

Albumin, 0.2 g	Balance, centigram
Bromine water, Br_2, saturated solution, 3 mL	Beral-type pipets, graduated, 7
Cholesterol, 0.2 g	Erlenmeyer flasks, 125-mL, 2
Coconut oil, 4 mL	Funnel and filter paper, large
Corn oil, 4 mL	Mortar and pestle
Hexane, C_6H_{14}, 30 mL	Spatula
Olive oil, 4 mL	Test tubes, small, 5
Peanuts, raw, 3–4	Test tube rack
Sudan III solution, 0.5% in alcohol, 2 mL	Wash bottle
Water, distilled or deionized	Wax or other marking pencil

Safety Precautions

Bromine water is toxic by inhalation and ingestion and is a skin irritant. Work with bromine water in an operating fume hood only. Avoid breathing bromine vapor. Sudan III solution is an alcohol-based solution and is a flammable liquid. Hexane is a flammable liquid and a dangerous fire risk. Do not allow any flames in the laboratory during this activity. Avoid exposure of all chemicals to eyes and skin. Some students may experience a food allergy to peanuts. Do not use peanuts in this experiment if any students are allergic to peanuts. Food-grade items that have been brought into the lab are considered laboratory chemicals and are for lab use only. Do not taste or ingest any materials in the chemistry lab. Wear chemical splash goggles, chemical-resistant gloves, and a chemical-resistant apron. Wash hands thoroughly with soap and water before leaving the laboratory.

Skim milk is produced commercially by using a centrifuge to separate the less dense cream from the more dense milk. The cream is then literally "skimmed off."

Procedure

Part A. Solubility of Lipids

1. Label a set of five test tubes 1–5 and place them in a test tube rack.

2. Use a Beral-type pipet or a spatula (in the case of solids) to add 10 drops of liquid or a small pinch of solid, respectively, to the appropriate test tube, as follows:

Test Tube	1	2	3	4	5
Sample	Coconut oil	Corn oil	Olive oil	Cholesterol	Albumin

3. Use a graduated Beral-type pipet to add 2 mL of water to each test tube. Shake or swirl the test tubes to mix the contents.

4. After 2 minutes, observe each mixture carefully. Has the sample dissolved in the water or are two separate layers evident in the tube?

5. Record the solubility results in Data Table A. Describe each sample as *soluble* S, if it dissolves completely; *slightly soluble* SS, if if dissolves partially; or *insoluble* IS if it does not appear to dissolve at all.

6. Add 5 drops of Sudan III staining solution to each test tube. Record the color and appearance of each test solution in Data Table A.

7. Rinse the contents of the test tubes down the drain with excess water. Wash the test tubes and rinse them with distilled or deionized water.

8. Relabel the tubes 1–5, if necessary.

9. Repeat step 2 to prepare a set of test tubes 1–5 for analysis.

10. Use a graduated Beral-type pipet to add 2 mL of hexane to each test tube. Swirl the test tubes to mix the contents.

11. After 2 minutes, observe each mixture carefully. Record the solubility results in Data Table A. Describe each sample as *soluble* S, if it dissolves completely; *slightly soluble* SS, if it dissolves partially; or *insoluble* IS if it does not appear to dissolve at all.

12. Pour the contents of the test tubes into a beaker marked "Organic Waste."

13. Wash the test tubes and rinse them with distilled or deionized water.

Part B. Test for Unsaturation.

Carry out these experiments in an operating fume HOOD only!

14. Relabel three test tubes 1–3, if necessary, and place them in a test-tube rack.

15. Use a Beral-type pipet to add 1 mL of the appropriate liquid to each test tube, as follows:

Test Tube	1	2	3
Sample	Coconut oil	Corn oil	Olive oil

Teacher Notes

16. Add 1 mL of bromine water to each sample. *Caution:* Handle with care! Do not breathe bromine vapor. Do not allow bromine to come in contact with skin or clothing.

17. Gently swirl each test tube once and replace it in the test tube rack.

18. After 2 minutes, observe the appearance of each bromine test mixture. Answer the questions in Part B of the Properties of Lipids Data Sheet.

19. Working in the **hood**, carefully pour the contents of the test tubes into a beaker marked "Bromine Waste." Rinse them once with distilled or deionized water and add the rinse water to the waste beaker.

Part C. Extraction of Peanut Oil

20. Obtain about 3–4 raw peanuts. Break the peanut shells and remove most of the thin skin covering the peanuts.

21. Transfer the shelled peanuts to a clean and dry mortar. Mash the peanuts with a pestle for at least 3–5 minutes to obtain finely divided, ground peanuts.

22. Tare (zero) a 125-mL Erlenmeyer flask and transfer about 3 g of ground peanuts to the flask. Record the exact mass of peanuts used in Data Table C.

23. Pour approximately 20 mL of hexane into the Erlenmeyer flask and stopper the flask.

24. Swirl the flask periodically over the next 15 minutes to mix the contents.

25. While the solvent extraction is proceeding, set up a filter funnel for gravity filtration. Measure and record the mass of a piece of filter paper and place it in a funnel set up over a second Erlenmeyer flask.

26. After 15 minutes, swirl the contents of the extraction flask and pour the contents into the funnel. Use a spatula to transfer any remaining solid from the Erlenmeyer flask to the funnel.

27. When the filtration is complete, carefully remove the filter paper containing the peanut residue from the funnel and place it in a secure location. Write your name on the paper and allow the residue to dry overnight.

28. Mass the peanut residue and filter paper. Describe the appearance of the residue and report its mass in Data Table C.

Name: _____

Class/Lab Period: _____

Properties of Lipids

Data Table A. *Solubility of Lipids*

Solubility Test	Test Tube and Sample				
	1	**2**	**3**	**4**	**5**
	Coconut Oil	Corn Oil	Olive Oil	Cholesterol	Albumin
Water					
Effect of Sudan III					
Hexane					

Part B. Test for Unsaturation

1. What is the initial color of bromine water?

   _____

2. What is the initial appearance of samples 1–3 immediately after the addition of bromine water?

   _____

   _____

   _____

3. Describe the final appearance of test samples 1–3 after two minutes.

   _____

   _____

   _____

4. Which seed oils reacted with bromine?

   _____

   _____

   _____

Properties of Lipids

Data Table C. *Extraction of Peanut Oil*

Mass of shelled peanuts	
Mass of filter paper	
Mass of filter paper + peanut residue	
Mass of peanut residue	
Appearance of peanut residue	

Post-Lab Questions *(Use a separate sheet of paper to answer the following questions.)*

1. Does the solubility behavior of the test samples in Part A fit the pattern predicted based on the definition of lipids? Explain. Which samples in Part A are lipids?

2. Compare the effect of Sudan III on solid and liquid lipids.

3. Sudan III stain can be used to identify fat storage granules in a seed. Discuss how this might be done and what might be observed.

4. The following information was obtained from the nutritional labels of various seed oils. Do the results of the bromine test for unsaturation agree with the information provided on the food labels?

	Coconut Oil	Corn Oil	Olive Oil
Total fat	14 g	14 g	14 g
Saturated fat	13 g	2 g	2 g
Polyunsaturated fat	0 g	4 g	1 g
Monounsaturated fat	1 g	8 g	11 g

5. How could the bromine test be modified to rank different seed oils with respect to the amount of unsaturation in each?

6. Using the mass of peanuts before and after extraction of "peanut oil," calculate the percentage of fat in peanuts.

7. The nutritional label for peanuts lists the following information. How does the experimental value for the percentage of fat in peanuts (Question #6) compare with that reported on the nutritional label?

Peanuts, 1 oz.	28.4 g
Total Fat	14 g

8. Describe some ways the extraction procedure in Part C could be improved to obtain better agreement between the experimental and known values of the amount of fat in peanuts.

See the CRC Handbook of Chemistry and Physics *for the iodine numbers of a variety of fats and oils. The iodine number gives a general indication of the degree of unsaturation in fats and oils.*

Teacher's Notes
Properties of Lipids

Master Materials List *(for a class of 30 students working in pairs)*

Albumin, 8 g	Balances, centigram, 3
Bromine water, 60 mL*	Beral-type pipets, graduated, 105
Cholesterol, 8 g	Erlenmeyer flasks, 125-mL, 30
Coconut oil, 75 mL	Funnels and filter paper, large (12.5-cm), 15
Corn oil, 75 mL	Mortar and pestles, 15
Hexane, C_6H_{14}, 500 mL§	Test tube racks, 15
Olive oil, 75 mL	Test tubes, 13 × 100 mm, 75
Peanuts, raw, 100 g	Wash bottles, 15
Sodium thiosulfate solution, $Na_2S_2O_3$, 50%, 250 mL†	Water, distilled or deionized
	Wax or other marking pencils

Sudan III solution, 0.5 % in alcohol, 50 mL

Organic waste beaker: Place a 250-mL beaker in the fume hood and label it "Organic Waste Beaker."

Bromine waste container: Place 100 mL of 50% sodium thiosulfate solution in a 500- or 600-mL beaker. Label the beaker "Bromine Waste."

*Bromine water does not store well. Prepare only as much solution as is needed for this lab. See the *Preparation of Solutions* section.

†For disposal of bromine water solutions. See the *Disposal* section.

§A mixture of n-hexane and other C_6H_{14} isomers.

Preparation of Solutions *(for a class of 30 students working in pairs)*

Bromine Water: Prepare 0.5 M sodium bromide and hydrochloric acid solutions. Combine 25 mL of 0.5 M NaBr and 25 mL of 0.5 M HCl in a dark glass (amber) bottle. Add 10 mL of 5% sodium hypochlorite solution (bleach) and gently swirl to mix the reactants. Cap the bottle and store it in a secure location. Carry out this procedure in an operating fume hood. Do **not** mix the HCl and NaOCl solutions directly in the absence of NaBr. *Note:* We recommend the purchase of the Bromine Water Kit (Flinn Catalog No. AP4502) for the safe preparation of bromine water in any amount in less than five minutes. Bromine water is very difficult to store because bromine vapor will escape from almost any container and begin to attack nearby metal surfaces. We highly recommend only making enough bromine water for your immediate needs. Do not store bromine water.

Sudan III solution, 0.5%: Warm 35 mL of 95% ethyl alcohol in a warm water bath. Do not boil. Add 0.25 g of Sudan III and stir to dissolve. Cool to room temperature and dilute to 50 mL with distilled or deionized water. Filter if needed.

Sodium Thiosulfate, 50%: Add 125 g of sodium thiosulfate pentahydrate (hypo, $Na_2S_2O_3 \cdot 5H_2O$) to 150 mL of distilled or deionized water. Stir to dissolve, then dilute to 250 mL. *Note:* The solution concentration is 50% by mass–volume.

Safety Precautions

Bromine water is toxic by inhalation and ingestion and is a skin irritant. Work with bromine water in an operating fume hood only. Avoid breathing the vapor. Do not allow bromine water to come in contact with skin or clothing. Sudan III solution is an alcohol-based solution and is a flammable liquid. Hexane is a a flammable liquid and a dangerous fire risk. Do not allow any flames in the laboratory during this activity. Avoid exposure of all chemicals to eyes and skin. Some students may experience a food allergy to peanuts. Do not use peanuts in this experiment if any students are allergic to peanuts. Food-grade items that have been brought into the lab are considered laboratory chemicals and are for lab use only. Do not taste or ingest any materials in the chemistry lab. Wear chemical splash goggles, chemical-resistant gloves, and a chemical-resistant apron. Consult appropriate Material Safety Data Sheets for additional safety, handling, and disposal information. Wash hands thoroughly with soap and water before leaving the laboratory.

Disposal

Consult your current *Flinn Scientific Catalog/Reference Manual* for both general guidelines and specific methods governing the disposal of laboratory waste. Lipids and proteins and Sudan III solution may be disposed of according to Flinn Suggested Disposal Method #26a or 26b, as appropriate. Hexane may be allowed to evaporate in a shallow pan placed in a fume hood, according to Flinn Suggested Disposal Method #18a. The organic waste solutions remaining after the hexane solubility test in Part A may be disposed of according to Flinn Suggested Disposal Method #18a or 18b. Excess bromine water should be reduced by reaction with sodium thiosulfate, according to Flinn Suggested Disposal Method #12a.

Lab Hints

- The experimental work for this lab can reasonably be completed in one 50-minute lab period. A few minutes of follow-up time are required the day after lab for students to mass the dry peanut residue from Part C.

- Preparation of bromine water requires about 15 minutes of prep time prior to class. This may be done 1–2 days before lab if the resulting solution is stored in a properly capped and labeled bottle in a secure location.

- To improve safety in the lab, set up two waste beakers in central locations for immediate collection of laboratory waste: (1) an "Organic Waste" beaker for the hexane solubility test samples in Part A; and (2) a "Bromine Waste" beaker filled with 100 mL of 50% sodium thiosulfate solution for the bromine test samples in Part B. Both waste beakers should be placed in the hood.

- Parts A, B, and C can be performed in any order. Consider staggering the starting points for different student groups—set up separate stations for Parts A, B, and C and have students rotate among the stations to complete the activity.

- Encourage students to label their pipets to avoid contamination and waste.

- Students sometimes become very indecisive about the solubility observations in Part A. Advise students to "call it as they see it."

- The test for unsaturation in Part B can be modified to rank different oils with respect to the relative amounts of saturated versus unsaturated fatty acids. Add bromine water dropwise to a given amount of oil until the red color of bromine persists and does not disappear. Consider this procedure only if there is enough time and sufficient hood space to avoid overcrowding.

- Are all nuts the same? Students might enjoy carrying out the extraction procedure in Part C with different types of nuts to compare how much fat they contain. Almonds, for instance, are a lower-calorie alternative to peanuts because they contain less fat. This could be a cooperative class exercise to compare the fat content in peanuts, walnuts, almonds, pecans, etc. Fresh nuts work better than dry roasted nuts.

- The extraction procedure in Part C can be extended to study the properties of peanut oil and the residual peanut "meal." Evaporate the solvent from the hexane/peanut oil extract and examine the behavior of peanut oil in the solubility and unsaturation tests. Dissolve the remaining solid peanut residue in water (most of it dissolves) and test it for protein using the Biuret Test (B0050, Biuret Test Solution) and for carbohydrate using Benedict's Test (B0171, Benedict's Qualitative Solution).

- One of the oldest chemical reactions known, dating back thousands of years, is the hydrolysis reaction of fats and oils with strong base to make soaps (fatty acid salts). The reaction is called saponification. An excellent kit is available from Flinn Scientific (Catalog No. AP4856, Soap-Making Kit) to allow students to make soap and study its properties.

Teaching Tips

- Encourage students to read the nutritional information labels on a variety of food items to learn about the distribution of saturated and unsaturated fats in different foods.

- The bromine test for unsaturation is similar to a test that is used by the food industry to characterize fats and oils. The iodine number, based on the reaction of ICl with fats and oils, is used as a quantitative measure of unsaturation. The greater the iodine number, the higher the percent unsaturation in a fat or oil. *Note:* ICl is used in the iodine number test because it is more reactive than I_2.

Hydrogenation of vegetable oils is also associated with another controversy in nutrition and food science—the presence of so-called "trans fats" or "trans fatty acids." The catalyst used in hydrogenation causes partial cis–trans isomerization of the double bonds in unsaturated fats.

Teacher Notes

Answers to Pre-Lab Questions *(Student answers will vary.)*

1. The $C≡C$ double bonds in unsaturated oils react with hydrogen to give "hydrogenated" compounds. What food items on your kitchen shelves have hydrogenated oils listed as a major ingredient? *Hint:* The snack food section is a good place to look.

   "Partially hydrogenated vegetable oil" is listed as a main ingredient in cookies, granola bars, crackers, margarine, potato chips, peanut butter, scalloped potato mixes, etc.

2. What should happen to the consistency of an oil when it is hydrogenated? What are the advantages of this change in consistency in food processing? What are the disadvantages?

   Hydrogenation converts unsaturated fatty acids to saturated derivatives and thus changes the consistency of an oil to a semi-solid form. The advantage of this in food processing is in improving the texture of food items with a high fat content. The chief disadvantage, of course, is that saturated fats are less healthy than unsaturated ones.

3. The fat content in milk can be measured by extraction with hexane. A 50.0-g sample of whole milk was mixed with hexane and gently stirred. Two liquid layers were obtained—a lower, "milky" layer and an upper layer of hexane. The layers were separated and the mass of the milky layer was found to be 47.8 g. Calculate the percentage of fat in whole milk.

   The difference in mass of the milk sample before and after extraction (50.0–47.8 g) equals the amount of fat contained in milk. Thus, 50.0 g of milk contains 2.2 g of fat. The percentage of fat in whole milk is calculated as follows:

   $$(2.2 \text{ g fat}/50.0 \text{ g milk}) \times 100\% = 4.4\% \text{ fat}$$

Sample Data

Student data will vary.

Data Table A. *Solubility of Lipids*

Solubility Test	Test Tube and Sample				
	1	**2**	**3**	**4**	**5**
	Coconut Oil	**Corn Oil**	**Olive Oil**	**Cholesterol**	**Albumin**
Water	IS	IS	IS	IS	S
Effect of Sudan III	oil layer is pink	oil layer is pink	oil layer is pink	solid is pink-speckled	solution is pale yellow
Hexane	S	S	S	S	SS

Part B. *Test for Unsaturation*

1. What is the initial color of bromine water?

   Dark orange-red.

2. What is the initial appearance of samples 1–3 immediately after the addition of bromine water?

   Each sample consists of two layers, a lower, dark orange layer due to bromine and an upper, pale yellow layer of vegetable oil.

3. Describe the final appearance of test samples 1–3 after two minutes.

   Sample 1 (coconut oil) did not change in appearance. The bromine color had not changed after three minutes. Samples 2 and 3 (corn oil and olive oil) changed significantly—the bromine color disappeared completely and the lower aqueous layer was left clear and colorless, while the top oil layer changed into a cloudy white emulsion.

4. Which seed oils reacted with bromine?

   Corn oil and olive oil.

Data Table C. *Extraction of Peanut Oil*

Mass of shelled peanuts	3.02 g
Mass of filter paper	1.33 g
Mass of filter paper + peanut residue	3.55 g
Mass of peanut residue	2.22 g
Appearance of peanut residue	pale, cream-colored powder

Answers to Post-Lab Questions *(Student answers will vary.)*

1. Does the solubility behavior of the test samples in Part A fit the pattern predicted based on the definition of lipids? Explain. Which samples in Part A are lipids?

   All of the lipid samples (coconut oil, corn oil, olive oil and cholesterol) were insoluble in water and soluble in hexane. Thus, lipids dissolve in a nonpolar organic solvent. The only non-lipid sample (albumin) had the opposite pattern—dissolving in water but not in hexane.

2. Compare the effect of Sudan III on solid and liquid lipids.

   Sudan III stained the liquid oil samples a uniform pink color, while the solid cholesterol sample absorbed the stain to give it a speckled appearance.

3. Sudan III stain can be used to identify fat storage granules in a seed. Discuss how this might be done and what might be observed.

   Cut a cross-section of a seed and add a few drops of Sudan III. Fat granules should appear a darker pink than the other parts of the seed.

4. The following information was obtained from the nutritional labels of various seed oils. Do the results of the bromine test for unsaturation agree with the information provided on the food labels?

	Coconut Oil	Corn Oil	Olive Oil
Total fat	14 g	14 g	14 g
Saturated fat	13 g	2 g	2 g
Polyunsaturated fat	0 g	4 g	1 g
Monounsaturated fat	1 g	8 g	11g

   The results of the bromine test agree with the information reported on the nutritional labels. Both corn oil and olive oil contain about 85% unsaturated fat and thus "decolorize" bromine. Coconut oil has less than 10% unsaturation and does not react with bromine.

5. How could the bromine test be modified to rank different seed oils with respect to the amount of unsaturation in each?

   Add bromine dropwise to oil until the added bromine does not react. Compare the color of the lower aqueous solution against a "blank" that does not react with bromine. Count the number of drops of bromine added until the color due to unreacted bromine persists and is the same as the blank. Rank oils based on the number of drops of bromine added—the more unsaturated the oil, the more bromine required for complete reaction.

6. Using the mass of peanuts before and after extraction of "peanut oil," calculate the percentage of fat in peanuts.

   Mass of "peanut oil" removed by extraction = 3.02 − 2.22 g = 0.80 g

   Percent fat in peanuts = (0.80 g/3.02 g) × 100% = 27%

7. The nutritional label for peanuts lists the following information. How does the experimental value for the percentage of fat in peanuts (Question #6) compare with that reported on the nutritional label?

Peanuts, 1 oz.	28.4 g
Total Fat	14 g

   Amount of fat listed on nutritional label for peanuts = (14 g/28.4 g) × 100% = 49%

   Only about half of the total fat was removed by the extraction procedure.

8. Describe some ways the extraction procedure in Part C could be improved to obtain better agreement between the experimental and known values of the amount of fat in peanuts.

   Increase the volume of solvent used, increase the length of time of the extraction procedure, carry out the extraction at a higher temperature, perform multiple extractions, stir the ground peanuts with the solvent continuously, grind the peanuts to a very fine powder before carrying out the extraction.

Membrane Diffusion
Dialysis Demonstration

Introduction

How does the membrane around a cell help to regulate the internal makeup of the cell? This demonstration compares the diffusion of small and large molecules across a semipermeable membrane to illustrate the process of diffusion in cells.

Concepts

- Semipermeable membrane

- Diffusion

- Dialysis

Materials

Beaker, 250-mL	Silver nitrate, $AgNO_3$, 0.1 M, 1 mL
Beral-type pipets, 2	Sodium chloride, NaCl, 1 M, 15 mL
Dialysis tubing, 6″ piece	Starch solution, 5%, 15 mL
Distilled or deionized water, 200 mL	String, 4″ piece, 2
Graduated cylinder, 10-mL	Test tube rack
Iodine solution, I_2/KI, 1 mL	Test tubes, 13 × 100 mm, 6

Safety Precautions

Silver nitrate solution is corrosive and can cause burns; it will stain skin and clothing. Iodine solution is irritating to eyes and skin. Avoid contact of all chemicals with eyes and skin. Wear chemical splash goggles, chemical-resistant gloves, and a chemical-resistant apron. Please consult current Material Safety Data Sheets for additional safety information.

Procedure

1. Obtain a 6″ length of pre-soaked dialysis tubing. The tubing is soaked in order to soften it for handling—twist the tubing between your thumb and index finger to make sure the tubing can be opened.

2. Close one end of the tube by tying a tight knot with a short piece of string about 1 cm from one end of the tubing. Be sure it is securely tied so that nothing can leak from the dialysis bag.

3. Carefully pour 10 mL of starch solution into the dialysis bag using a 10-mL graduated cylinder. Rinse the graduated cylinder, then pour 10 mL of sodium chloride solution into the dialysis bag.

4. Use another short piece of string to close the open end of the dialysis tubing. Tie it tight enough to prevent leakage.

5. Rinse the dialysis bag thoroughly under running water so that there is no starch or salt solution on the outside of the dialysis bag.

Dialysis tubing clamps that eliminate the need for tying the bags with string are available from Flinn Scientific (Catalog No. AP4349). See the Master Materials Guide (pp 70–72) for recommended solutions for this demonstration.

6. Place approximately 200 mL of distilled or deionized water into a clean 250-mL beaker. Place the dialysis bag into the beaker of water.

- 250-mL Beaker
- Salt-free Water
- Dialysis Bag (with starch and salt solution)
- String Seal

7. Set the beaker in a place where it can sit undisturbed for 10–15 minutes. Complete Steps 8 and 9 while waiting.

8. Label four test tubes 1–4 and place them in a test tube rack.

9. Prepare the test tubes as described below. Observe the tubes before and after the testing solutions (iodine and silver nitrate) have been added.

   #1   5 mL water + 3 drops iodine solution (control)

   #2   5 mL water + 3 drops silver nitrate solution (control)

   #3   5 mL starch solution + 3 drops iodine solution (positive test for starch)

   #4   5 mL salt solution + 3 drops silver nitrate solution (positive test for chloride ions)

10. After 10–15 minutes, use two clean test tubes to test the water in the beaker containing the dialysis bag. Prepare the test tubes as follows:

    #1   5 mL beaker water + 3 drops of iodine solution

    #2   5 mL beaker water + 3 drops of silver nitrate solution

11. Discuss the results of the control (step 9) and dialysis (step 10) experiments. Compare the permeability of the membrane with respect to salt and starch.

Disposal

Please consult your current *Flinn Scientific Catalog/Reference Manual* for general guidelines and specific procedures governing the disposal of laboratory waste. The waste solutions may be disposed of down the drain with excess water according to Flinn Suggested Disposal Method #26b. Excess silver nitrate solution may be disposed of according to Flinn Suggested Disposal Method #11.

Tips

- Prepare 5% starch solution by heating 100 mL of distilled or deionized water to boiling. Remove the water from the heat and slowly sprinkle in 5 g of soluble starch while stirring. Heat for a few more minutes, with continued stirring, until the starch is dissolved.

- Test the tap water in your school to determine if it tests positive with silver nitrate for the presence of chloride ions. If the water tests positive, use distilled water or bottled water for the demonstration. If it tests negative, then tap water may be used.

- Cut the dialysis tubing into 6″ pieces and place in water to soak for about 15 minutes before use. To avoid a false positive test for starch in the beaker, make sure the dialysis tubing is securely tied and rinse the outside of the bag before placing it in the beaker.

- Silver nitrate and iodine are used as test solutions to detect the presence of sodium chloride and starch, respectively. Sodium chloride and silver nitrate react to form silver chloride, a milky-white precipitate. Iodine reacts with starch to form a purple-blue complex.

$$NaCl(aq) + AgNO_3(aq) \longrightarrow AgCl(s) + NaNO_3(aq)$$

$$starch(aq) + I_2(aq) \longrightarrow starch–iodine\ complex$$

Discussion

Diffusion (the random movement of molecules) is one of the key processes involved in the movement of materials throughout living systems. Diffusion can be effectively demonstrated by observing the action of a drop of dye in a glass of water. The water molecules in the water are in constant, random motion. If a drop of blue food dye is added to the glass, the dye begins to slowly diffuse throughout the water. The individual dye molecules disperse and, compelled by collisions with the moving water molecules, eventually become evenly distributed throughout the solution. The water in the glass will finally appear to be a uniform shade of lighter blue. The movement of individual molecules in this system is indeed random. The net movement of the dye molecules is directional, in the sense that they initially move from a region where they are highly concentrated (the dye droplet) to where they are less concentrated (the surrounding water). Thus diffusion (the term usually refers to the net movement of molecules) is said to occur down a concentration gradient— from a region of high concentration to a region of lower concentration.

Cells are enclosed within membranes that regulate the movement of materials into and out of the cell. Some molecules are small enough to pass freely in and out of the cell through the membrane. Other molecules cannot diffuse freely through the membrane. Thus, cell membranes are said to be selectively permeable or semipermeable. If we envision the membrane material as being porous, then it is easy to imagine that some molecules are small enough to fit through the pores while others are too large. Dialysis is the separation of smaller molecules from larger molecules by selective diffusion through a semipermeable membrane.

Dialysis tubing is used in this activity to simulate a cell membrane. It is made of selectively permeable cellulose that is perforated with microscopic pores. The pores are small enough that the tubing can be used to model the behavior of the cell membrane with respect to the sizes of molecules that will (or will not) diffuse through them. Sodium chloride dissociates in aqueous solution to produce sodium (Na^+) and chloride (Cl^-) ions that are small enough to pass through the membrane. Starch is a very large, high-molecular weight carbohydrate (polysaccharide) that is too large to diffuse through the membrane.

Demonstrate the effect of temperature on the rate of diffusion by conducting the dialysis experiment at three different temperatures. Use beakers containing ice water, room-temperature water, and hot water.

Glucose Fermentation
Metabolism Demonstration

Introduction

Interest in fermentation, the breakdown of sugar to alcohol, dates far back in human history. In 1860, Louis Pasteur showed that fermentation involves a living process carried out by yeast and bacteria. The overall products of the fermentation of glucose, the main carbohydrate in fruits and grains, are ethyl alcohol and carbon dioxide. Many different intermediate products may also be formed depending on reaction conditions. In this demonstration, the production of carbon dioxide and ethyl alcohol will be illustrated. A redox indicator, resazurin, will also be used to highlight the conditions present during the fermentation process.

Concepts

- Carbohydrates
- Oxidation
- Glucose
- Fermentation

Materials

Bromthymol blue solution, 0.04%, 2 mL
Glucose (dextrose), 75 g
Lugol's iodine solution, 1 mL*
Resazurin solution, 0.1%, 1 mL
Sodium hydroxide solution, 1 M, 1 mL*
Yeast, active dry, 1.5 g
Water, distilled, 150 mL
Balance
Condenser*
*Optional equipment for Iodoform Test

Erlenmeyer flask, 125-mL
Hot plate*
Pipets, Beral-type, 2
Plastic or latex tubing, 12″
Stopper, one-hole, with glass tube
Test tube, 15 × 125 mm
Test tube rack
Watch glass*

Safety Precautions

Wear chemical splash goggles, chemical-resistant gloves, and a chemical-resistant apron. Please consult current Material Safety Data Sheets for additional safety information.

Procedure

1. Place about 10 mL of distilled or deionized water in the test tube, followed by 2 mL of bromthymol blue indicator solution. Stir well to mix. *Note:* The indicator solution in the test tube should be green or blue to start. If not, add one drop of base to the solution.

2. Place 75 mL of distilled water into a 125-mL Erlenmeyer flask. Add 7.5 g of glucose (also called dextrose) to the flask and swirl the contents until all the glucose has dissolved.

3. Add 1.5 g of active dry yeast to the glucose solution in the Erlenmeyer flask and swirl until the yeast is uniformly mixed in the solution.

4. Add 10 drops of resazurin to the solution in the flask.

5. Use a one-holed stopper, a glass tube, and flexible plastic or latex tubing to complete the fermentation setup shown in Figure 1. The plastic or latex tubing should be below the surface of the bromthymol blue indicator solution in the test tube.

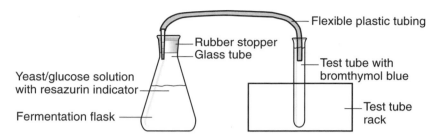

Figure 1. Alcoholic Fermentation Setup

6. Note the initial color of the solution in both the flask (purple) and the test tube (green).

7. Let the fermentation apparatus sit undisturbed for 10–15 minutes. Record the color changes of the solutions in both the flask and test tube.

8. *(Optional)* Save the solution in the flask to perform the iodoform test for alcohol (Steps 9–13).

9. When the yeast activity in the fermentation flask has ceased and bubbling has stopped (24 hours), decant the supernatant (liquid layer) into a clean flask.

10. Connect the new flask to a condenser and heat the liquid to distill the ethyl alcohol.

11. Catch the first portion of the distillate and place 2–3 mL of distillate onto a watch glass.

12. Add 3–5 drops of sodium hydroxide solution to the distillate, followed by 3–5 drops of Lugol's iodine solution.

13. Heat the watch glass gently on a hot plate and note the formation of yellow iodoform crystals. *Note:* Iodoform (CHI_3) is toxic and has a disagreeable, pungent odor. Do not handle the solid. Work in a fume hood or well-ventilated area.

Disposal

Please consult your current *Flinn Scientific Catalog/Reference Manual* for general guidelines and specific procedures governing the disposal of laboratory waste. Small amounts of iodoform may be disposed of following Flinn Suggested Disposal Method #26a. The contents of the fermentation flask may be disposed down the drain with excess water according to Flinn Suggested Disposal Method #26b.

The use of distilled or deionized water is strongly recommended. Tap water may contain impurities that will inhibit or kill the yeast cells.

Tips

- Consult your school regulations before performing the optional distillation/iodoform test for alcohol. Some schools have a zero-tolerence policy for alcohol and drugs that precludes their use even in scientific experiments.

- This demonstration may be set up for every class—each setup will be at a different stage throughout the school day. Alternatively, one demonstration may be set up and monitored by various classes throughout the day.

- The reactions in this demonstration are temperature sensitive. If the surroundings are warm, the gas production will occur within minutes.

Discussion

The fermentation reaction of glucose by yeast produces ethyl alcohol and carbon dioxide (Equation 1). This is an anaerobic process—it occurs in the absence of oxygen.

$$C_6H_{12}O_6(aq) \longrightarrow 2CH_3CH_2OH(aq) + 2CO_2(g) \qquad \textit{Equation 1}$$
$$\text{glucose} \qquad\qquad \text{ethyl alcohol} \qquad \text{carbon dioxide}$$

Resazurin is a redox indicator that is commonly used to show the depletion of oxygen in solution by living microbes. When microbes are present, the resazurin goes through a series of color changes indicating the oxygen condition of the solution. In this demonstration, the yeast suspension quickly consumes the available oxygen and produces the anaerobic conditions required for alcohol fermentation. The solution in the flask changes color from an initial deep purple to pink, peach, orange, and finally colorless as the oxygen concentration is depleted.

As carbon dioxide is produced in the fermentation process, it bubbles through the tube and into the bromthymol blue solution in the test tube. As the CO_2 bubbles through the solution, the pH gradually decreases and the color of bromthymol blue, an acid–base indicator, changes from green to yellow.

The optional iodoform test provides a qualitative test for the other product resulting from yeast fermentation, ethyl alcohol. Ethyl alcohol reacts with iodine in the presence of sodium hydroxide to give iodoform, CHI_3. See Equation 2.

$$CH_3CH_2OH + 4I_2 + 6NaOH \longrightarrow CHI_3 + HCOONa + 5NaI + 5H_2O \quad \textit{Equation 2}$$

Bromthymol blue is an acid–base indicator. It is blue when the pH >7.6, yellow when the pH <6.0, and green in the transition range, pH 6.0–7.6.

Teacher Notes

Lactose Intolerance
Enzyme Digestion Demonstration

Introduction

Intestinal gas—not a very pleasant thought! It can be a big problem for those individuals who lack the enzymes to digest certain foods. Milk and dairy products, for example, cause problems for many people who lack the enzyme required to digest lactose, the main carbohydrate in milk. This demonstration illustrates the use of a commercial enzyme product called Lactaid™ as an aid in milk digestion.

Concepts

- Enzyme
- Disaccharide
- Monosaccharide

Materials

Balloons, 4

Lactose, 10 g

Glucose, 5 g

Galactose, 5 g

Yeast, active dry, 12 g

Lactaid™, 1/2 tablet

Erlenmeyer flasks, 125-mL, 4

Mortar and pestle

Water bath, 35–40 °C

Water, 500 mL

Safety Precautions

Wear chemical splash goggles, chemical-resistant gloves, and a chemical-resistant apron. Please consult current Material Safety Data Sheets for additional safety information.

Procedure

1. Prepare a warm water bath (35–40 °C) for use in the demonstration—a clear glass baking pan with hot tap water will work nicely. The water bath is needed to allow the dramatic balloon expansion to occur within a typical 50-minute class period.

2. Weigh out the dry ingredients prior to the demonstration, and grind ½ tablet of Lactaid in a mortar with a pestle.

3. Review the summary diagram of the demonstration setup shown in Figure 1.

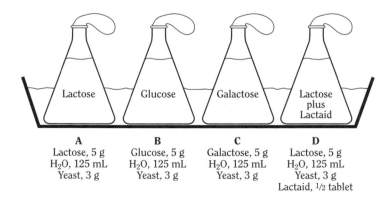

A	B	C	D
Lactose, 5 g	Glucose, 5 g	Galactose, 5 g	Lactose, 5 g
H₂O, 125 mL	H₂O, 125 mL	H₂O, 125 mL	H₂O, 125 mL
Yeast, 3 g	Yeast, 3 g	Yeast, 3 g	Yeast, 3 g
			Lactaid, ½ tablet

Figure 1. Summary of Demonstration Setup

4. Clearly label each flask as shown in Figure 1.

5. Place 5 g of the appropriate dry sugar into each flask, as shown in Figure 1.

6. Add about 110 mL of warm (30–35 °C) tap water to the sugar in each flask and swirl each flask until all the sugar has dissolved.

7. Add 3 g of yeast to each flask. Swirl each flask gently to mix in the yeast.

8. Place a balloon securely over the lip of each flask. *Note:* Be sure each balloon is flexible and not stuck together. Inflate each balloon at least once before placing it on the flask.

9. Place all four flasks in the water bath (35–40 °C).

10. Observe the flasks for 15–30 minutes, checking for the production of gas as observed in the balloons. Discuss the results and the effectiveness of Lactaid™ in the experiment.

Disposal

Please consult your current *Flinn Scientific Catalog/ Reference Manual* for general guidelines and specific procedures governing the disposal of laboratory waste. The waste solutions may be disposed of down the drain with excess water according to Flinn Suggested Disposal Method #26b.

Tips

- Yeast lacks the enzymes necessary to digest lactose or galactose, but it does contain the enzyme needed to digest glucose. See Figure 2 for a summary of the demonstration results.

- The balloon in Flask B inflates rapidly due to the production of carbon dioxide from digestion of glucose.

- In Flask D, Lactaid effectively breaks down the lactose to give glucose and galactose. The yeast then further digests the glucose to give alcohol and carbon dioxide, which causes the balloon to inflate.

- Flasks A and C should reveal no activity and the balloons should remain uninflated for the duration of the demonstration.

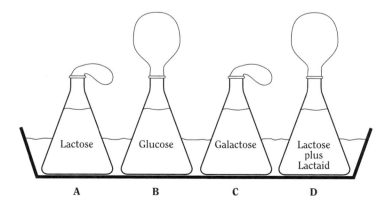

Figure 2. Summary of Demonstration Results

- This demonstration can serve as a springboard for lab project extensions. How much lactose can a Lactaid tablet digest? How long does the Lactaid remain active? What factors influence the rate of the reaction?

Discussion

Lactose (also called milk sugar) is the principal carbohydrate in milk. It is a *disaccharide*, which means that it is composed of two simple sugars or monosaccharides—glucose and galactose. Some individuals produce insufficient quantities of *lactase,* the enzyme required to break the bond between the two monosaccharide units in lactose. The condition in which lactase is not produced so that lactose cannot be broken down into the two simpler sugars is referred to as *lactose intolerance*. With this condition the lactose from milk and various other milk products remains undigested and causes an increase in the osmotic pressure in the intestinal contents. Consequently, water is "drawn" from the tissues into the intestine. At the same time, intestinal bacteria may act upon the undigested lactose and produce organic acids and gases. As a result, the person may feel bloated and suffer from intestinal cramps, diarrhea, and gas.

Researchers at Lactaid, Inc. discovered a way to mass-produce the enzyme lactase. Lactase converts lactose into glucose and galactose, both easily digestible monosaccharides. The mass-produced lactase is formulated into the tablet product—Lactaid™. The enzyme units in Lactaid are eaten by the lactose-intolerant person before or simultaneous to ingesting lactose. The Lactaid breaks down the lactose and then the person's system can utilize the resulting glucose and galactose.

An external test for the effectiveness of Lactaid is to find an organism that will only digest the simple sugars and not lactose. In this experiment, yeast is used as a test organism.

Yeast does not produce lactase and therefore cannot digest lactose. (It is lactose intolerant!) Yeast does, however, digest glucose very efficiently. When it digests glucose, the yeast breaks down the glucose and produces carbon dioxide gas as a waste product (Equation 1). This production of gas can be easily monitored. The absence or presence of gas production is used as evidence of digestion by the yeast.

$$C_6H_{12}O_6(aq) \longrightarrow 2CH_3CH_2OH(aq) + 2CO_2(g) + energy \qquad \textit{Equation 1}$$
$$\text{glucose}$$

See "Introduction to Carbohydrates" in the Experiments *section of this lab manual for a more complete discussion of mono- and disaccharides. The structures of glucose and lactose are illustrated on page 12.*

Amino Acid Fingerprints
Ninhydrin Demonstration

Introduction

Detectives use ninhydrin to reveal fingerprints left at crime scenes. Ninhydrin reacts with amino acids found in the natural oils on our skin to produce a purple product. The intensity of the color may also be used as quantitative test for the amount of amino acids in a sample.

Concepts

- Amino acids
- Ninhydrin
- Forensic chemistry

Materials

Blotting paper

Ninhydrin solution in alcohol, 0.5%

Hot plate or high temperature hair dryer

Spray bottle

Inkpad, water-soluble

Gloves

Safety Precautions

Work in a fume hood or well-ventilated area. Ninhydrin is an irritant, biologically active, and is usually dissolved in an alcohol solvent. Alcohol solvents are flammable liquids; keep away from flames or other sources of ignition. Ninhydrin will stain skin. Wear chemical splash goggles, chemical-resistant gloves, and a chemical-resistant apron. Please consult current Material Salety Data Sheet for additional safety information.

Procedure

1. Ask students to make a set of fingerprints along the top half of a sheet of blotting paper.

2. Holding the paper in a gloved hand, spray the paper with ninhydrin solution. Allow the paper to dry for a few minutes before picking it up. This will prevent the fingerprints from "running."

3. When the paper is dry enough that the solution will not run, pick up the paper and allow it to air dry completely using a fanning motion.

4. When dry, hold the filter paper above a heat source such as a hot plate. Hold the paper about 10 cm above the heat source to prevent scorching. *Note:* It takes about 2–3 minutes of heating over the low setting of the hot plate for the fingerprints to appear.

5. A set of purple prints or spots will soon appear.

6. Have students produce a corresponding set of fingerprints along the bottom edge of the blotting paper using a water-soluble inkpad.

7. Compare the two sets of fingerprints.

Disposal

Blotting paper may be disposed of in the trash following Flinn Suggested Disposal Method #26a. The ninhydrin solution may be stored for future use.

Some people have very pronounced fingerprints, while others may leave only smudges. To get better fingerprints, have students run their fingers through their hair several times to pick up more oil.

Teacher Notes

Tips

- Ninhydrin solution may be prepared in ethyl, isopropyl, or butyl alcohol. For convenience, we recommend the purchase of ready-made ninhydrin solution in butyl alcohol (Flinn Catalog No. N0039).

- All amino acids and most proteins give positive results in the ninhydrin test. The ninhydrin test is commonly used to detect amino acid "spots" in the separation of amino acids by paper chromatography. The purple color does not develop until the amino acid–ninhydrin mixture is heated.

- In forensic chemistry, ninhydrin is most often used to detect latent fingerprints left behind on porous surfaces—cloth, paper, and cardboard. Prints on hard, nonabsorbent surfaces, such as mirrors, tile, or glass, are detected using fingerprint powders such as graphite or aluminum dust.

- A series of reference and control tests may be performed to demonstrate the positive reaction of ninhydrin with amino acids and proteins. Prepare a set of 1% reference solutions containing amino acids, such as phenylalanine and tyrosine, and proteins, such as albumin and gelatin. Add 2 mL of ninhydrin solution to 1 mL of each reference sample in a test tube, and heat the solutions at 75–80 °C for 3–5 minutes. If desired, these positive reference tests may be compared against a series of controls that give negative results. Water, salt and sugar should all test negative with ninhydrin.

Discussion

Ninhydrin is a pale yellow solid. It reacts with amino groups in amino acids and proteins to produce a purple product (Equation 1). The reaction is very slow in the absence of heat—heat is therefore used to speed up the reaction.

ninhydrin   amino acid   *Equation 1*

purple product $+ \; CO_2 + RCHO + 3H_2O$

Latent fingerprints are composed of several chemicals that are naturally present in skin oils or released through the pores of the skin via perspiration. Some of the chemicals in fingerprints include sodium chloride, amino acids, glucose, lactic acid, and ammonia.

pH and Protein Solubility
A Reversible Demonstration

Introduction

Any change in the pH of a protein's environment will cause observable changes in the solubility of the protein. These changes reflect changes in the three-dimensional structure of the protein. The effect of pH on protein solubility explains why most enzymes function well at an optimum pH, and why their activity decreases substantially at pH values other than the optimum. This demonstration examines the effect of pH on the solubility and structure of casein, a milk protein.

Concepts

- Protein
- Solubility
- Isoelectric point
- pH

Materials

Casein, 2 g	Beaker, 600-mL
Hydrochloric acid, HCl, 3 M, 50 mL	Magnetic stirrer and stirring bar
Sodium hydroxide, NaOH, 0.01 M, 250 mL	Pipets, Beral-type, 2
Sodium hydroxide, NaOH, 3 M, 50 mL	Spatula
Universal indicator (optional), 2 mL	Weighing dish
Balance	

Procedure

1. Place a beaker containing 250 mL of 0.01 M sodium hydroxide and a stirring bar on a magnetic stirrer. Stir at moderate speed. Add 2 g of casein and stir to dissolve. The solution will be slightly cloudy or translucent.

2. *(Optional)* Add 1–2 mL of universal indicator to monitor pH, if desired. Consult the universal indicator color chart to follow the solubility of casein as a function of pH.

3. Add 3 M hydrochloric acid one pipet-full at a time using a Beral-type pipet. A white precipitate will form after the addition of about 10–15 mL of hydrochloric acid. The pH at which the precipitate appears is 4–5.

4. Continue adding hydrochloric acid with stirring. The precipitate will redissolve after the addition of another 15–20 mL of acid (pH <2).

5. Reverse the process by adding 3 M sodium hydroxide with stirring. The protein will precipitate out again after about 20–30 mL of sodium hydroxide have been added. This occurs as the pH increases through the 4–10 range.

6. Continue adding sodium hydroxide dropwise with stirring. The precipitate will redissolve as more base is added and the pH increases above pH 10–12.

Do not allow the precipitated casein to stand too long. Upon prolonged standing, the precipitated protein will clump together and become denatured. Once the protein is denatured, it may not redissolve again as excess acid or base is added.

Teacher Notes

Disposal

Please consult your current *Flinn Scientific Catalog/Reference Manual* for general guidelines and specific procedures governing the disposal of laboratory waste. The casein solution may be stored at basic pH for several months. Alternatively, the solution may be rinsed down the drain with excess water according to Flinn Suggested Disposal Method #26b.

Discussion

Casein is the principal protein in milk, comprising 80% of the total protein content in milk. Casein is a phosphoprotein—it contains a large number of phosphate groups attached to the amino acid side chains in its polypeptide structure. The negatively charged phosphate groups are balanced by positive calcium ions and are responsible for the high nutritional calcium content in milk. Casein is almost completely insoluble in water at neutral pH (pH = 7).

Casein, like other proteins, is an ionic species containing amino groups and carboxyl groups on its terminal amino acids. It also contains a variety of other acidic and basic groups on the side chains of its non-terminal amino acids. The effect of pH on the solubility of casein reflects the ionization of the acidic and basic groups in its structure.

At high pH, casein will have a net negative charge due to ionization of its acidic side chains ($—CO_2^-$). Because casein is ionized at high pH values, it is soluble in dilute sodium hydroxide solution.

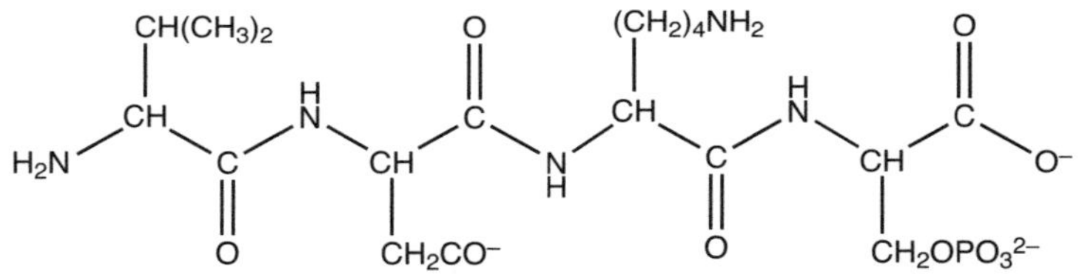

At low pH, casein will have a net positive charge due to protonation of its basic side chains ($—NH_3^+$). Because casein is ionized at low pH values, casein is also soluble in strongly acidic solutions.

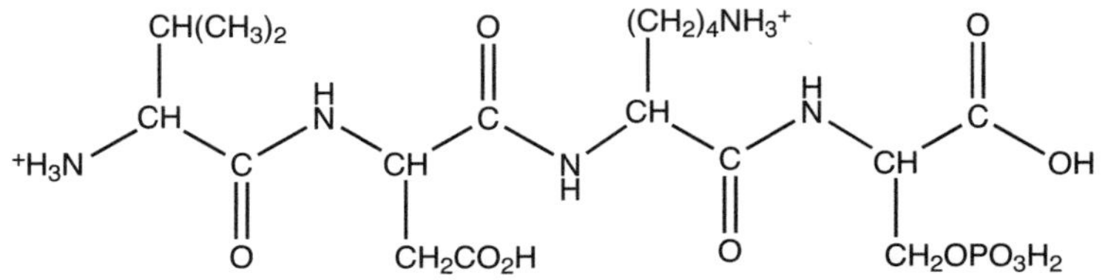

At intermediate pH values, casein will contain an equal number of positively and negatively charged groups and the protein will have a net charge of zero. Casein is insoluble in neutral solutions because it is not charged under these conditions.

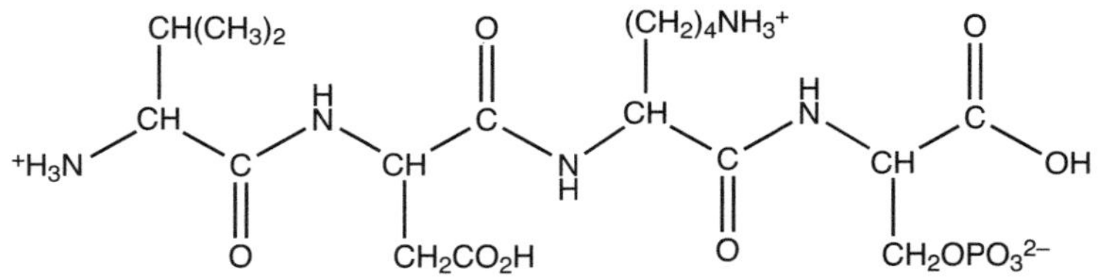

The structures shown here do not represent the actual structure of casein. They are shown only to indicate the general types of acidic and basic groups that may be present in proteins.

The solubility of a protein is usually at a minimum at its isoelectric point. The isoelectric point is defined as the pH at which a protein has a net charge of zero. For casein, due to the attached phosphate groups, the isoelectric point is close to pH = 4.

65

pH and Protein Solubility

Safety and Disposal Guidelines

Safety Guidelines

Teachers owe their students a duty of care to protect them from harm and to take reasonable precautions to prevent accidents from occurring. A teacher's duty of care includes the following:

• Supervising students in the classroom.

• Providing adequate instructions for students to perform the tasks required of them.

• Warning students of the possible dangers involved in performing the activity.

• Providing safe facilities and equipment for the performance of the activity.

• Maintaining laboratory equipment in proper working order.

Safety Contract

The first step in creating a safe laboratory environment is to develop a safety contract that describes the rules of the laboratory for your students. Before a student ever sets foot in a laboratory, the safety contract should be reviewed and then signed by the student and a parent or guardian. Please contact Flinn Scientific at 800-452-1261 or visit the Flinn Website at www.flinnsci.com to request a free copy of the Flinn Scientific Safety Contract.

To fulfill your duty of care, observe the following guidelines:

1. **Be prepared.** Practice all experiments and demonstrations beforehand. Never perform a lab activity if you have not tested it, if you do not understand it, or if you do not have the resources to perform it safely.

2. **Set a good example.** The teacher is the most visible and important role model. Wear your safety goggles whenever you are working in the lab, even (or especially) when class is not in session. Students learn from your good example—whether you are preparing reagents, testing a procedure, or performing a demonstration.

3. **Maintain a safe lab environment.** Provide high-quality goggles that offer adequate protection and are comfortable to wear. Make sure there is proper safety equipment in the laboratory and that it is maintained in good working order. Inspect all safety equipment on a regular basis to ensure its readiness.

4. **Start with safety.** Incorporate safety into each laboratory exercise. Begin each lab period with a discussion of the properties of the chemicals or procedures used in the experiment and any special precautions—including goggle use—that must be observed. Pre-lab assignments are an ideal mechanism to ensure that students are prepared for lab and understand the safety precautions. Record all safety instruction in your lesson plan.

5. **Proper instruction.** Demonstrate new or unusual laboratory procedures before every activity. Instruct students on the safe way to handle chemicals, glassware, and equipment.

6. **Supervision.** Never leave students unattended—always provide adequate supervision. Work with school administrators to make sure that class size does not exceed the capacity of the room or your ability to maintain a safe lab environment. Be prepared and alert to what students are doing so that you can prevent accidents before they happen.

7. **Understand your resources.** Know yourself, your students, and your resources. Use discretion in choosing experiments and demonstrations that match your background and fit within the knowledge and skill level of your students and the resources of your classroom. You are the best judge of what will work or not. Do not perform any activities that you feel are unsafe, that you are uncomfortable performing, or that you do not have the proper equipment for.

Safety Precautions

Specific safety precautions have been written for every experiment and demonstration in this book. The safety information describes the hazardous nature of each chemical and the specific precautions that must be followed to avoid exposure or accidents. The safety section also alerts you to potential dangers in the procedure or techniques. Regardless of what lab program you use, it is important to maintain a library of current Material Safety Data Sheets for all chemicals in your inventory. Please consult current MSDS for additional safety, handling, and disposal information.

Disposal Procedures

The disposal procedures included in this book are based on the Suggested Laboratory Chemical Disposal Procedures found in the *Flinn Scientific Catalog/Reference Manual*. The disposal procedures are only suggestions—do not use these procedures without first consulting with your local government regulatory officials.

Many of the experiments and demonstrations produce small volumes of aqueous solutions that can be flushed down the drain with excess water. Do not use this procedure if your drains empty into groundwater through a septic system or into a storm sewer. Local regulations may be more strict on drain disposal than the practices suggested in this book and in the *Flinn Scientific Catalog/Reference Manual*. You must determine what types of disposal procedures are permitted in your area—contact your local authorities.

Any suggested disposal method that includes "discard in the trash" requires your active attention and involvement. Make sure that the material is no longer reactive, is placed in a suitable container (plastic bag or bottle), and is in accordance with local landfill regulations. Please do not inadvertently perform any extra "demonstrations" due to unpredictable chemical reactions occurring in your trash can. Think before you throw!

Finally, please read all the narratives before you attempt any Suggested Laboratory Chemical Disposal Procedure found in your current *Flinn Scientific Catalog/Reference Manual*.

Flinn Scientific is your most trusted and reliable source of reference, safety, and disposal information for all chemicals used in the high school science lab. To request a complimentary copy of the most recent *Flinn Scientific Catalog/Reference Manual,* call us at 800-452-1261 or visit our Web site at www.flinnsci.com.

Experiments and Demonstrations

Content Standards	Introduction to Carbohydrates	Identifying Proteins and Amino Acids	Physical Properties of Proteins	Properties of Lipids	Membrane Diffusion	Glucose Fermentation	Lactose Intolerance	Amino Acid Fingerprints	pH and Protein Solubility
Unifying Concepts and Processes									
Systems, order, and organization	✓	✓			✓	✓			
Evidence, models, and explanation	✓	✓	✓	✓	✓	✓	✓	✓	✓
Constancy, change, and measurement			✓		✓				✓
Evolution and equilibrium					✓				✓
Form and function	✓	✓	✓	✓	✓			✓	✓
Science as Inquiry									
Identify questions and concepts that guide scientific investigation	✓	✓	✓	✓	✓		✓		✓
Design and conduct scientific investigations	✓	✓	✓	✓	✓	✓	✓		✓
Use technology and mathematics to improve scientific investigations									
Formulate and revise scientific explanations and models using logic and evidence	✓	✓	✓	✓	✓		✓		
Recognize and analyze alternative explanations and models									
Communicate and defend a scientific argument									
Understanding scientific inquiry	✓	✓	✓	✓	✓		✓		
Physical Science									
Structure of atoms									
Structure and properties of matter	✓	✓	✓	✓					✓
Chemical reactions	✓	✓	✓	✓	✓	✓	✓	✓	✓
Motions and forces									
Conservation of energy and the increase in disorder					✓				
Interactions of energy and matter									

Experiments and Demonstrations

Content Standards *(continued)*

	Introduction to Carbohydrates	Identifying Proteins and Amino Acids	Physical Properties of Proteins	Properties of Lipids	Membrane Diffusion	Glucose Fermentation	Lactose Intolerance	Amino Acid Fingerprints	pH and Protein Solubility
Science and Technology									
Identify a problem or design an opportunity									
Propose designs and choose between alternative solutions									
Implement a proposed solution									
Evaluate the solution and its consequences									
Communicate the problem, process, and solution									
Understand science and technology							✓	✓	
Science in Personal and Social Perspectives									
Personal and community health			✓			✓			
Population growth									
Natural resources									
Environmental quality									
Natural and human-induced hazards									
Science and technology in local, national, and global challenges									
History and Nature of Science									
Science as a human endeavor									
Nature of scientific knowledge	✓	✓	✓	✓	✓	✓			✓
Historical perspectives	✓	✓	✓	✓		✓			

(for a class of 30 students working in pairs)   **Experiments and Demonstrations**

Chemicals	Flinn Scientific Catalog No.	Introduction to Carbohydrates	Identifying Proteins and Amino Acids	Physical Properties of Proteins	Properties of Lipids	Membrane Diffusion	Glucose Fermentation	Lactose Intolerance	Amino Acid Fingerprints	pH and Protein Solubility
Albumin	A0258		2 g	8 g	8 g					
Ammonium sulfate	A0062			280 g						
α-Naphthol	N0067	0.05 g								
Arginine	A0308		1 g							
Barfoed's reagent	B0061	250 mL								
Benedict's solution	B0015	200 mL								
Biuret test solution	B0051		150 mL							
Bromine Water Kit	AP4502				1					
Bromthymol blue solution, 0.04%	B0173						2 mL			
Casein	C0043		2 g	1 g						2 g
Cholesterol	C0179				8 g					
Coconut oil	C0074				75 mL					
Corn oil	C0090				75 mL					
Cupric sulfate pentahydrate	C0102			2.5 g						
Cysteine	C0420		1 g							
Ethyl alcohol, 95%	E0009		50 mL		35 mL					
Fructose (Levulose)	L0020	1 g								
Galactose	G0041							5 g		
Gelatin	G0037		2 g	1 g						
Glucose (Dextrose)	D0005	1 g					7.5 g	5 g		
Hexane	H0054				500 mL					
Hydrochloric acid solution, 3 M	H0034	125 mL		100 mL						50 mL
Iodine	I0006	0.15 g								
Iodine–potassium iodide solution	I0038					1 mL				
Isopropyl alcohol	I0019			50 mL						
Lactose	L0002	1 g						10 g		
Lugol's iodine solution	I0036						1 mL			
Ninhydrin solution, 0.5%	N0039								100 mL	
Nitric acid, 3M	N0049		150 mL							
Olive oil	Q0004				75 mL					
Potassium iodide	P0066	0.75 g								
Resazurin solution, 1%	R0013						1 mL*			
Resorcinol	R0015	0.13 g								
Silver nitrate	S0274			0.85 g						